Science and
Technology in China

Science and Technology in China

Implications and Lessons for India

Edited by

Maharajakrishna Rasgotra

 www.sagepublications.com
Los Angeles • London • New Delhi • Singapore • Washington DC

First published in 2013 by

 SAGE Publications India Pvt Ltd
B1/I-1 Mohan Cooperative Industrial Area
Mathura Road, New Delhi 110 044, India
www.sagepub.in

SAGE Publications Inc
2455 Teller Road
Thousand Oaks, California 91320, USA

SAGE Publications Ltd
1 Oliver's Yard, 55 City Road
London EC1Y 1SP, United Kingdom

SAGE Publications Asia-Pacific Pte Ltd
33 Pekin Street
#02-01 Far East Square
Singapore 048763

Published by Vivek Mehra for SAGE Publications India Pvt Ltd, Phototypeset in 10/13 Berkeley by Tantla Composition Pvt Ltd, Chandigarh and printed at Saurabh Printers Pvt Ltd.

Library of Congress Cataloging-in-Publication Data

Science and technology in China: implications and lessons for India/edited by Maharajakrishna Rasgotra.
     pages cm
  Includes bibliographical references and index.
  1. Science and state—China.   2. Technology and state—China.
  3. Technology indicators—China.   4. Technology indicators—India.
  5. Technological forecasting—India.   I. Rasgotra, M., editor of compilation.
  Q127.C5S275   509.51—dc23   2013   2013019799

ISBN: 978-81-321-1312-6 (HB)

The SAGE Team: Rudra Narayan, Dhurjjati Sarma, Anju Saxena and Rajinder Kaur

To the memory of Jawaharlal Nehru—
scientist, statesman and first Prime Minister of India—
who foresaw the extraordinary role of science and
technology in transforming India

Thank you for choosing a SAGE product! If you have any comment,
observation or feedback, I would like to personally hear from you.
Please write to me at <u>contactceo@sagepub.in</u>

—Vivek Mehra, Managing Director and CEO,
SAGE Publications India Pvt. Ltd, New Delhi

Bulk Sales

SAGE India offers special discounts for purchase of books in bulk.
We also make available special imprints and excerpts from our
books on demand.

For orders and enquiries, write to us at

Marketing Department
SAGE Publications India Pvt. Ltd
B1/I-1, Mohan Cooperative Industrial Area
Mathura Road, Post Bag 7
New Delhi 110044, India
E-mail us at <u>marketing@sagepub.in</u>

Get to know more about SAGE, be invited to SAGE events, get on
our mailing list. Write today to <u>marketing@sagepub.in</u>

This book is also available as an e-book.

Contents

List of Tables

List of Figures

Introduction

Maharajakrishna Rasgotra

In November–December 1951, I was accompanying a Chinese cultural delegation—the first ever to visit a foreign country—headed by China's Vice Minister for Culture, Mr Ting Si Lin, on an extensive tour of India. Jawaharlal Nehru, Prime Minister of India, was also on a tour of central and south India. Travelling in those days was mostly by road, and the paths of the two motorcades crossed at a couple of places—once near Bhopal in Madhya Pradesh, and then further down the road in Andhra Pradesh. In Madhya Pradesh, at a small wayside gathering of a few hundred unlettered villagers and their children, Nehru was telling his enraptured audience to send their children to schools, even if the schools were some distance away from their villages: government will open new schools, colleges and universities, and they should educate their children in schools and colleges, and teach them science and maths, and other subjects too; India needs doctors, engineers and scientists in very large numbers to become a strong and prosperous nation. "In the modern world", he emphasized, "science is very important, a nation can become strong and make progress only if it is good in science and maths and engineering."

The memory of the Bengal famine and the looming threat of food shortage still haunted India, and Nehru also spoke to the villagers about what his government was doing to improve agriculture and enhance the production of food grains. He talked to them about big dams and canals, small irrigation projects, agricultural universities and new fertilizer factories—the scientific infrastructure of a green revolution—his government was laying which would, in the short course of a decade, transform productivity in India.

Barely five years from independence, Nehru's government had rehabilitated some 11 million non-Muslim refugees who had poured into India from Pakistan and moved on to the tasks of rebuilding an India battered by famine, poverty and disease, and of reviving a moribund economy ravaged by two centuries of colonial exploitation. Nehru's emphasis on higher education in general, and on education in science and engineering etc., in particular, was his mantra for India's development and for bringing the country into the modern age. But while there was a great deal to do, there was acute shortage of resources in the country. What Nehru did with great foresight and success was to lay the foundations in all the various areas of India's need for growth and advancement, which the succeeding generations could build upon. Science and technology received his special attention, for he wanted India to develop into a country with a 'scientific temper'.[1]

In 1951–52, despite the paucity of resources and other problems confronting India, the country's mood was one of self-discovery, enthusiasm and hope. With the mushrooming of schools and colleges, industrial estates, factories big and small, the planning of new townships, roads, dams and irrigation projects, and the opening of new institutes of science and engineering, there was a palpable feeling of forward movement. The country was abuzz with activity and a throbbing feeling of self-renewal. The Chinese visitors were impressed with what they saw in different regions of India. They evinced special interest in the size and the low estimated cost of the Bhakra-Nangal project. The Nilokheri community development project near Delhi seemed to rouse their curiosity; they wanted to know how many Nilokheris were there all over India! They had found particularly intriguing Nehru speaking about the importance of scientific and technical education to villagers; they asked: did his audience include highly educated people? The leader of the delegation was impressed, in particular, with Nehru's emphasis on science and technology as the tools for India's reconstruction and advancement.

Nehru's interest in science and technology was of long standing because during his years of study in England, and in his several visits to European countries and Russia later, he had observed that science and technology were the two great engines of the West's progress and power. He was acquainted with the work of eminent Indian scientists of the pre-independence era—C.V. Raman, Jagdish Chandra Bose, Srinivasa

Ramanujan, Visveswaraya and others. On another occasion, when I was asked to accompany King Mahendra of Nepal on a two-week tour of India, Nehru had instructed me to take the king and his entourage to meet the Nobel laureate C.V. Raman in his lab in Bangalore and request him to show the King the 'Raman Effect', which was, in fact, the important discovery of laser. As soon as the Indian awards and honours were instituted, Raman was one of the three illustrious personages to be honoured by Nehru's government with the highest award of Bharat Ratna (Jewel of India) in 1954—the sage statesman Rajagopalachari and India's philosopher-president Radhakrishnan were the other two. This was Nehru's way of emphasing the importance of science in the national life of the India of his dreams. I cannot help noting, that while the number of Bharat Ratnas has swelled to 41, after 1954 only one other scientist, Dr A.P.J. Abdul Kalam (1997) has been admitted to that charmed circle of bejeweled eminence. I can think of at least six scientists, past and present, deserving of that high honour.

However, despite the honour conferred on Raman, little further development of laser for industrial and other uses took place in India in the early years of independence. In 1980, when I visited a laser laboratory in France, its director said to me that the world owed the discovery of laser to India, and he wondered why India itself had not exploited its potential for industrial and other uses. The only explanation I could offer was that before independence the British government of India had little interest in giving encouragement and support to Indian science, and after independence India's limited resources were absorbed by other priorities. In fact, C.V. Raman, in his speech accepting the Noble Prize, had castigated the imperial powers for their contemptuous disregard of poor countries' science. Of course, thanks to Nehru, things had changed after Independence, and the director recognized that Indian science had made impressive strides, especially in the nuclear field. An even warmer tribute to Indian science came from President Giscard d'Estaing of France, when I presented my credentials to him in February 1979. He remarked that with several Indian scientists winning Nobel Prizes, India's proper place was among the developed, not the developing countries.

I imagine what had happened was that with the invention and the military use of the nuclear weapon on the eve of India's independence, science and technology had taken a long leap in a new direction signifying radical changes in the very concepts of power and security. Naturally, therefore, Nehru's attention was centred on all the various aspects of that new development and its implications for India's security. Therefore, soon after taking the reins of government, Nehru had invited three important contemporary scientists to enhance the reach and scope of Indian science, including, especially, its nuclear dimension. The most influential of the three, and closest to Nehru personally, was Homi Jehangir Bhabha, who founded and developed India's nuclear programme and recruited many young scientists to work with him to create a new and vibrant scientific establishment. P.C. Mahalanobis, Director of the Indian Statistical Institute, helped Nehru's government in the formulation of economic and industrial policies and in planning. Shanti Swarup Bhatnagar was put in-charge of the Council of Scientific and Industrial Research (CSIR) and the Department of Science and Industrial Research. Bhatnagar's main contribution was to set up a slew of laboratories for scientific and industrial research in different parts of the country.

In 1961, when the Department of Atomic Energy was well established, Nehru asked it to begin work in space research. In the following year, Homi Bhabha set up a Committee for Space Research with Vikram Sarabhai as its Chairman, who conceived and launched, in 1963, the first sounding rocket from Thumba. That was the beginning of India's complex and highly successful space programme which over the years has made impressive advances in the exploration of space and in the civilian uses of the knowledge gained from the programme. It needs now to be linked more closely with India's defence.

By the mid-1980s, India's science establishment had grown to a point that we often boasted, not without some justification, to have the third largest scientific manpower in the world after the USA and the USSR. That, however, did not remain the case for long.

China had begun creating an industrial base to manufacture major arms in the late 1950s and within a decade its nuclear programme had surpassed India's nuclear programme, which pre-dated China's. The wars of 1962, 1965 and 1971, fractious domestic politics, and the sanctions imposed on India by the USA and by the Nuclear Suppliers Group after India's nuclear explosion of 1974 took a heavy toll on the country's resources leading to a weakening of the focus on science and technology, precisely when it needed an organizational overhaul and expansion. In comparison, under Deng Xiao Ping, China's science and technology (S&T) establishment, whether in the number of new colleges and universities, technology institutes and science laboratories, or in the enrolment in science subjects in schools and higher institutions, underwent a massive expansion and upgrading in a well-planned and integrated framework combining corresponding expansion and diversification of China's manufacturing industry, the export earnings of which paid for the modernization and enlargement of the country's science and technology network.

In India, we have neglected manufacturing for export. The bulk of our exports to China comprises iron ore. But perhaps the worst example of this neglect is aviation, a critical factor in the modern world's civil convenience and military power. In both China and India, aviation is a public sector undertaking. Professor Roddam Narasimha points out in the chapter on aviation and aeronautics in China that AVIC (Aviation Industry Corporation of China) has 400,000 employees compared with around 40,000 employees in HAL, the only significant aircraft industry in India. China now manufactures its own civil and military aircraft and also exports them; India continues to be dependent on imports for both.

Nehru's successors did not build sufficiently on his legacy, and because of the absence of opportunities for creative and rewarding employment in India, Indian scientists, engineers and doctors have been migrating to the US and European countries. The situation deteriorated further in the decade-and-half following Rajiv Gandhi's defeat in the 1989 general election. This was the time of the development of special economic zones on the shores of the South China Sea and when China was beginning to lure back to China, non-resident Chinese scientists from the USA and other western countries in large numbers to man and develop hundreds of new S&T institutes on advanced U.S. models. In India, on the other hand, due to the absence of an imaginative policy to integrate science, technology, R&D, innovation and industry for a clear national objective, even the capacities created earlier were allowed to stagnate. India's progress in S&T also continued to be hampered by the absence of a large-scale manufacturing industry. S&T and industry stimulate and sustain mutual growth, but in India, unlike in China, industry and S&T operate in their separate domains, and the industry that does exist in the country contributes little to R&D. There is a policy lacuna in this domain which needs to be addressed. Prime Minister Manmohan Singh's government has adopted a manufacturing policy in 2011, but a proper technology policy is still missing. Our defence industry still functions in isolation from the civilian economy and the S&T sector: they need to be closely integrated for a more rapid advancement of all three sectors. These pressing issues do not appear to have received much attention in the centenary session of the Indian Science Congress at Kolkata in December 2012.

On the other hand, while Prime Minister Manmohan Singh announced the new science policy of his government for the next decade and spoke of his government having "invested like never before in science",[2] the scientists complained of red-tapism continuing to obstruct research and demoralise them, of the inadequacy of actual investment in R&D at below one per cent of GDP, despite the goal set in 2003 of investment of two per cent of GDP which the new policy had now re-set for the next decade. There is criticism of the new science policy on other counts too, but it must be said to the credit of Prime Minister Singh that during his nine years at the helm he has established 24 new central universities, which are now functional, and 16 institutes of innovation, established in 2009, are beginning to put their act together. After Rajiv Gandhi, who demitted office in 1989, Dr Manmohan Singh is the only Prime Minister to give due importance to science and higher education. His policies in this regard will help strengthen the pursuit by India's future rulers of a practice begun by Nehru of extending patronage and support to S&T and R&D.

Patronage of science and technology is not a part of our millennial tradition. The *navratnas* of India's royal courts from time immemorial to the last ruling *maharaja* or *nawab* were saints and seers, philosophers, musicians and humourists, political counsellors, warriors, poets and panegyrists, and scribes and sycophants of varying talent. Prime Minister Manmohan Singh has done much to strengthen a tradition begun by Nehru and sustained by Indira Gandhi and Rajiv Gandhi.

Competence and competitiveness of a nation in science and technology not only aid its economic and social development, they also enhance its standing and influence in the world. More important, a nation's scientific and technological strength confers on it critical military advantages in ensuring its own security and in acquiring a role in world affairs. All this is particularly relevant for a country like India whose size, resource base, human talent, history, geography and power potential have ordained for it an important peace and security role beyond its national confines. Information Technology, in particular, has changed the nature of warfare, and this is an area in which China has left India a very long way behind in every dimension, from education and research to supercomputing. According to Professor Balakrishnan, the stunning growth of IT in China has enabled the country to improve its capability and sophistication in warfare to more or less match the USA and Russia!

With focus only on software and services, IT development in India is flawed and incomplete. Ninety percent of India's telecommunications and IT hardware is imported. The threat to security inherent in this situation apart, the economy of a large country like India, with massive unemployment and other shortcomings, cannot sustain itself and thrive on services and software alone.

My main concern in all this is not that China will launch another war against India; rather, the more worrying implication for India is about this nation of 1.2 billion people not being able to command the prestige and influence it deserves, and to play its proper role in ensuring peace, security and stability in its region and beyond and leave an imprint of its own on the evolving global civilization.

Professor Ramamurthy tells us that the quality of Indian S&T is comparable to China's, 'investment per researcher is comparable; research output per researcher is comparable; we do not simply have as many numbers of researchers on the work bench as China'. This theme of the modesty of scale of India's S&T recurs in most of the pages that follow, and the major cause for it is the inadequacy of investment in this all-important enterprise. The scale of China's undertakings in all the various S&T areas is mind-boggling; what is even more impressive is the vast, albeit somewhat wasteful, organization China has created in integrating its numerous S&T and R&D and policy-making establishments and a huge number of production enterprises into one seamless giant comprising 11,000 large and small industries employing 700,000 scientific and technological personnel. They are now trying to remove waste and tighten the organization.

However, there is no over-claim in what Professor Ramamurthy says about the basic soundness and competence of Indian S&T: clearly, there are achievements to India's credit in areas of agriculture, genetic drugs, wind energy, photovoltaics and fast breeder technology, and even in the quality and efficiency of steel-making technology, which China has not yet matched in performance. There is no over-claim even in Professor Parthasarathi's assertion that even Chinese S&T has not come close to matching the innovative excellence of some of the products of Sam Pitroda's C-Dot and Ashok Jhunjhunwala's Wireless in Local Loop (WiLL).

We do not have to compete with or match the scale of China's S&T or the size of its massive financial investment in S&T and engage in a fruitless race with China. Let us hope, instead, that China–India relations will, before long, reach a stage when the two countries can cooperate in S&T for the benefit of their huge populations and the larger world community. But in the context of the needs of India's development and security, there is no reason, now, why the scale and scope of India's S&T cannot be transformed. For, India no longer suffers from the crippling shortages of human and financial resources which characterized the first three or four decades after independence. But we need to define a clear goal and put together the resources to pursue it with determination. And this is a task for India's political elite.

Dr Kharbanda, while acknowledging India's successes in scientific and technological domains, says India is caught in a paradox of too much democracy! He recommends a big political push for S&T in India to isolate it from the stresses of the country's increasingly partisan politics. He probably had in mind the experience of India–US civil nuclear deal, one of Prime Minster Manmohan Singh's more significant achievements for the country, which was so stoutly opposed by the Opposition in Parliament for no good reason. To further compound matters, in an act of national self-denial, the Parliament adopted legislation (the Nuclear Liability Act) which has, for the time being at least, prevented India from reaping the gains of the deal.

S&T does not necessarily flourish only in a highly centralized and regimented system of governance. If there is the necessary political consensus on national needs and goals, democracy, even a rumbustious one like ours, is no bar for a country's rapid rise in science and technology. After all, advances in S&T, greater by far than China's or of the former USSR, were made, and continue to be made, in democratic countries like the USA, Britain, Germany, France, Japan and Israel.

There is some criticism, perhaps not really justified, in this book of China's seemingly unseemly methods: for example, reverse engineering. China is none the worse for its policies and methods; the world has acclaimed China's success and is eager to profit from it. As for reverse engineering, as Smitha Purushottam says, everyone has been there, including India. Nothing prevented India from reverse-engineering and manufacturing at home at least some of the wide variety of arms we have been importing for decades. The problem lies in the absence of the ambition and the urge to make this huge country self-reliant in meeting its indispensable needs of the wherewithal of defence through indigenous production. Our military, with the exception to some extent of the Indian Navy, prefers periodic acquisition of 'state-of-the-art' weaponry from foreign vendors, without the requisite effort to promote manufacturing their requirements in India. And for some reason, our ruling elite seems complicit in this mistaken course to equip India's armed forces. Economic development in all major countries—the US, Russia and China—was dominated, in the early periods of their growth, by defence technologies.

China is India's largest and most powerful neighbour. We need to know this great country better and improve our knowledge and understanding of its policies; and for this purpose India needs its own un-biased literature on China. With that objective, four years ago, we started a seminar series on China at the Observer Research Foundation (ORF) where I was President in-charge of research activity till recently. Only Indians conversant with China's society, culture and internal and external policies were invited to present papers and take part in discussion in those seminars. Observer Research Foundation has published five books compiled from that seminar series on China's Politics, Society and Culture; Economy and Environment; Contemporary China and the World; China in South Asia and A Net Assessment of Present-Day China. This is the sixth volume in the series and, arguably, a seminal one, in the sense that it may lead to more studies on aspects of China's S&T—and also India's—not dealt with here.

This book is not a comprehensive, comparative study of S&T in China and India, though unavoidably, comparisons occur here and there in every chapter. Ashok Parthasarathi's chapter, in particular, does intentionally attempt a comparative appreciation of the quality and quantum of progress achieved in the two countries in half-a-dozen critical industrial sectors. This is of interest as a comparative examination of this kind does not seem to have been attempted before. My hope is that more books will follow not only on the branches of China's S&T not dealt with in this volume, but also on S&T in India. In a knowledge society we claim to be building in India, it is necessary to disseminate information on these subjects in the general public. Hopefully, government may also take note of what our scientists are saying about why India has fallen so far behind China in S&T from the lead position it held till the mid-1980s.

The scientists who took part in the seminar, of which this book is the outcome, are persons of high eminence in their respective domains and in great demand internationally and within India. It was not easy to assemble them and keep them at ORF in Delhi for two whole days. Some of them, good personal friends of mine, had worked with me in the National Security Advisory Board and in other fora. I simply have no words to adequately express my appreciation and gratitude to them for their contributions. The award, earlier this month, of Padma Vibhushan to Roddam Narasimha gave me a long-awaited thrill. Surely, it is time for Indian Science to be given another place or two in the haloed Bharat Ratna ranks.

I am grieved to record the passing of Dr Kharbanda whom I came to admire deeply during our interactions in our seminar. His sudden death on 15 January came as a deep personal shock to me; and, of course, it is a serious loss to Indian science. He will be missed by the science community for long.

I owe much to Ambassador K. Raghunath, a respected former colleague of mine in the Indian Foreign Service, who was of immense help in organizing and conducting the seminar. I am also thankful to Dr K. Yhome, Fellow at ORF and my Secretary, P.J. Biju, for their uncomplaining labours at their computer desks, for much repetitive work required to prepare the manuscript for publication.

The writing style of this book is not typically academic; nor is it uniform in style, unlike the assiduously written single-author books.

This book is the outcome of oral PowerPoint presentations, followed by vigorous discussion among the authors and a few additional invitees. The transcripts were vetted by me and sent to the authors for revision, modification and improvements as they thought fit. We did not think it prudent to prescribe any wordage limit for them. In editing the final texts received from them, I took special care not to tamper, beyond a point, with the writing style of the authors so as to retain the spontaneity and authenticity of thought, feeling and expression. Here and there a colloquial expression may intrude upon the reader's attention, but he may even enjoy the experience of the author actually trying to converse with him.

Notes and References

1   At the 40th session of the Indian Science Congress at Lucknow on 2 January 1953 Nehru said:

> We should fashion a society where the real scientist will play a more important part in developing [and] helping that society to function, and in promoting that scientific temper, or even temper, which has become quite essential not only for progress but even if we have to survive.

Nehru spoke on the subject in the same vein at the 1948 Science Congress and on numerous other occasions.

2   The Prime Minister said this in his inaugural address at the Indian Science Congress at Kolkata in December 2012. The *Hindustan Times* and other Indian dailies carried fairly detailed reports of the Prime Minister's address.

1

Science and Technology in China

Implications and Lessons for India: An Overview

V.S. Ramamurthy

As regards a comparison of Science and Technology (S&T) in India and China, the first question we need to ask is how much the two countries are investing in research and development.

When I received the invitation to give an overview of 'Science & Technology in China: Implications & Lessons for India' at the Observer Research Foundation's Seminar on the subject, my mind went back about 40–50 years which were the late years of my education and early years of my professional career in the Atomic Energy Establishment, Trombay, the present Bhabha Atomic Research Centre, Mumbai. At that time, we all used to have Camlin pens and it was supposed to be a good pen. We would like to have a Parker pen but it was too expensive to buy. Some of my friends used to have Parker pens made in China, and many of us used to say, "This is no good because it doesn't write as good as a Camlin." We were indeed happy to have the Camlin made in India in place of the Parker made in China, because in the 1960s and 1970s the perception of China was that it made cheap consumer goods of questionable quality and by what one might call unfair means. For example, it used to be said that the Parker Company was asked to set up a plant in China, and then, one day, they were asked to pack up and go, and the company became a Chinese company making China-made Parker pens.

Some 20 years later, I was in United States with my daughter to attend my grandson's first birth anniversary. At the end of the day, we had about 100 packets, gifts of various kinds, and the usual ritual is to sit and start unwrapping them. Every one of those gifts, except the one which I had carried from India, my daughter told me were one dollar items made in China. To buy a fighter plane in those days one had to go to an American producer, but most of the small day-to-day things were all made in China, and they were in the market at acceptable prices and of acceptable quality: though China is now making fighter jet also, the situation in regard to consumer items of daily use is not very different.

Ten years later in 2008, I saw a newspaper report saying that a subsidiary of the Chinese Meteorological Department had outbid Bharat Electronics Limited (BEL) to supply Doppler Weather Radars to India. India had a plan to have something like 55 Doppler Weather Radars and this was the first order for about 12 units. It struck me as odd, because in the late 1990's, the India Meteorological Department (IMD) had funded Bharat Electronics and the Department of Space to design, build and install a Doppler Weather Radar, which is functioning in Sriharikota since 1998–99. Subsequently, we also bought couple of Doppler Weather Radars from a German company. What China did in 2008 was to outbid both of them! How is a Chinese company, which has already had the technology from none other than the Indian Department of Space, able to outbid an Indian supplier? I asked my friends in the IMD what had happened and why did it happen? They said the Chinese product was cheaper and the software was better. This was, indeed strange, because in one case we were paying to BEL and the Department of Space, which means from one arm of the government to another, and in the other case, we were giving it to China! Where, then, is the question of which is expensive and which is cheaper? As for our software not being as good, we are supposed to be the software leader in the world. More important, what happened in the last 10 years is that no development took place using the functioning unit. Within two years of that happening, questions were being asked whether by putting the Chinese Doppler Weather Radar we were compromising the security of our air space.

One message is clear: the Chinese competition is no longer limited to small, low value items; it extends to high value, high technology products as well. With good technology at competitive prices, China is already in. I will leave the security discussions to a little later. China is clearly moving up the value chain in the global market, starting with very poor quality products to high technology products in about 30 to 40 years. In the early years, China was obviously concentrating on simple manufacturing and marketing with not very high technology inputs. In fact, we used to talk about notorious violations of IP in China in the early years. But, today, this is not necessarily true. For example, the Doppler Weather Radars themselves were designed and manufactured jointly with Lockheed Martin. So, obviously it is not a stolen technology; it is something which has been acquired through proper channels, and the so-called violator of IP has turned out to be second only to the US in patent filing.

China is a major patent filer today in the world. Clearly there is a change taking place in China's technological capability, manufacturing capability, and to some extent, in the marketing capability. What is contributing to this phenomenon is something which needs to be understood.

During the same period, India had not been doing badly either. We started off with practically no infrastructure on anything and we built, over a period of a few decades, a good educational infrastructure, a research infrastructure and an industry infrastructure. And in selected areas, like agriculture or generic drugs, we have achievements to our credit. We went on from ship-to-mouth existence to food surplus by using technology (Green Revolution). We were net importers of generic drugs in the 1950s and the 1960s, and today we are net exporter of generic drugs. So clearly, we have not been doing badly; but from the early years up to the 1990s the emphasis was on import substitution and reverse engineering. So obviously, we were also violators of IP; IP protection was not one of the high priority items at that time.

In the 1990s things changed with economic liberalization, and our initial reaction was we were going to be completely overrun by the external markets and we would be finished. But we survived the competition—and there was global competitiveness in selected sectors like automobiles and vaccines! So clearly, we too have been moving ahead. We are moving forward, but the question which we keep asking ourselves is that at the rate at which China is growing in the technology chain, and the rate at which India is going on the same path, can India survive Chinese competition in the coming years? Just as in the US, our markets are also full of 'made-in China' products, cost and quality competitive, often displacing many indigenous products. The famous Aligarh locks have disappeared from the Indian marketplace making way for the made-in-China locks? We are seeing more and more silk saris made in China, better and cheaper than saris made in Mysore or Kancheepuram? Where are we heading? That is the question which we have to keep asking ourselves.

I shall not dwell on how China is doing, or India is doing, in space technology, nuclear technology and other strategic sectors, such as defence production, because these are going to be dealt with by other contributors and also because in such areas each country has its own ambitions, its own strategy, and it is very difficult to compare one with another. But I do want to touch on one aspect of strategy—not strategy in the military's space but strategy in the marketplace. And it is about the Rare Earths Story.

The term 'rare earths' refers to a group of seventeen elements in the periodic table, specifically, the 15 Lanthanides plus Scandium and Yttrium. This group of elements holds unique physical, chemical, magnetic, electrical and luminescence properties, and combined with other materials they can also alter their physical and chemical properties in unique ways. The rare earths magnets, which are much smaller than the standard iron magnets have very wide applications in the automobile industry and in renewable energy industry. Wherever there is a motor, there is a magnet, and wherever there is magnet, you want to make it more powerful and more compact. Today, China produces over 90 per cent of the low-value and 99 per cent of the high-value rare earths oxides for world consumption and controls 97 per cent of the global rare earths market. But rare earths resources of China account for only about 37 per cent. How it happened is a clear example of how China combines scientific and technological research, economics and strategic thinking very effectively.

Until the middle of the last century, most of the world's rare earths were sourced from sand deposits in India and Brazil, not from China at all. Through the 1950's, South Africa came into the market because there were some new findings of rare earths bearing monazites and then through the 1960s and 1980s the Mountain Pass Rare Earths in California was the leading producer in the United States. In 1986, for some reason which cannot be just without a basis, the Chinese government placed rare earths on the list of top secret national priorities, and in 1992, the Chinese Prime Minister informed the world: "The Middle East has oil, China has rare earths." This was a statement made by the Chinese Prime Minister.

A few years later in a policy decision in the late 1980s and 1990's, the President of China said to his own people: improve the development and applications of rare earths, and change the resource advantage into economic superiority. By 1997, the rare earths producer in the United States was forced to stop mining because of the increasing pricing pressure from China! The Chinese basically outbid the Americans in price. Of course, there was also help from the United States itself—environmental pressure from the State of California, leading to stoppage of all operations by 2002! In 2003, China acquired one of the most advanced rare earths magnetic facility in the United States, then closed it and transplanted the entire plant to China including its portfolio of patents in 2003. Now, while the company was being bought by China, they also trained the manpower so that the whole thing was manned by Chinese manpower. In less than 30 years China had made rare earths into a national monopoly!

The Chinese were also using a mixture of restrictive production, export policy, tax regime on rare earths, with the object, basically, of shifting more and more rare earths technology-dependent manufacturing facilities within the borders of China. So, one would never know when one would get the raw material. Or, not get it at all! So, put up the plant in China! Japan had a taste of this near monopoly policy, when Beijing halted shipments to Japan over a territorial dispute: it had nothing to do with anything other than a territorial dispute. Clearly, China has not only built the monopoly, it also knows how to twist things around in its favour in unrelated matters. Now, of course, efforts are being made in the US and elsewhere to catch up with China in this field, but this is not going to happen overnight, because the Americans had closed their factories, dismissed their manpower, and there is nothing left there of the earlier operations. They have to start all over again when 97 per cent of the market is controlled by China.

In the 1950's we were the major producer of rare earths. We actually stopped producing rare earths in 2004 due to lack of market competitiveness. In the 1990s, there was a proposal to have a titanium plant in India which was scuttled because it was not economically viable, because the cost of production was more than the cost of Chinese titanium. What is the manufacturing cost in China? God only knows! Their marketing is what we need to understand. They closed the titanium market last year, and I understand some product prices went up by a factor of 10. Talking about competitiveness in the market and taking a policy decision like closing that activity is not a great strategy. Of course, now we are trying to revive our capabilities to become competitive once again, both through indigenous efforts as well as with international co-operation—with Japan for example. Considering that India has a major stake in renewable energy and major ambitions in the automobile sectors both of which require rare earths, it is indeed surprising that we did not have a policy on rare earth minerals. But perhaps, it is not surprising; because we rarely have any strategy in the marketplace.

Are there lessons for India in all this? India and China are the two most populous countries of the world accounting for nearly one-third of the world population. Both countries, while having different political systems, have adopted science and technology for their development, both economic development and national security. They have chosen different paths and their achievements have been different.

As regards a comparison of Science and Technology (S&T) in India and China, the first question we need to ask is how much the two countries are investing in scientific research and development. We should begin with a comparison of India and China in terms of their research intensity. Table 1.1 gives some figures related to the research intensity for India, China and the US.

While India and China are comparable countries in terms of their population, in terms of GDP, our GDP is a little less than half of China's. What is even more important is that our gross expenditure on R&D (GERD) in PPP terms in dollars is only about one-fourth of what China is spending. Clearly, we are investing a good deal less than China. Of course, the GERD/GDP ratios of 1.4 for China and 0.8 for India themselves compare poorly with 2.7 in many of the technologically advanced countries. What is even more disturbing is that our 0.8 has remained 0.8 for the past about 20 years. China, on the other hand, started off with 0.8 and has been ramping up very fast. China wants

Table 1.1

Research Intensity

	World	USA	China	India
Population (millions)	6671	309 (4.6%)	1329 (19.9%)	1165 (17.5%)
GDP PPP$billions	66294	13741 (20.7%)	7103 (10.7%)	3100 (4.7%)
GERD PPP$billions	1146	373 (32.6%)	102 (8.9%)	25 (2.2%)
GERD/GDP	1.7	2.7	1.4	0.8
GERD per capita PPP$	172	1209	77	21
GERD per researcher PPP$thousands	159	244	72	127

Source: UNESCO Science Report 2010, 2007 Statistics.

to catch up with the advanced countries. Though in terms of per unit population, our investments are about one-fourth of China's, we are not doing badly in terms of expenditure per researcher. What it means is that the number of researchers in India is not as high as in China. These are the broad parameters.

How do the two countries compare in terms of their scientific outputs? What are the indicators that reflect the returns on the investments in R&D? The number of scientific publications and patents filed are some of the quantitative measures of research activity in a country. China beats India in terms of mere numbers, though both fall far below the US. On the other hand, it is somewhat comforting to note that the performances of the two countries are comparable when one looks at the scientific outputs per unit investment basis. In fact, the quality of scientific output in the two countries is also comparable based on the citations. Wherever India has participated in international mega science projects, Indian contributions have always been internationally competitive.

I would like to conclude as follows: In terms of the quality of output for the investments, it is not fair to say that our S&T system is poorer than China's S&T. Quality is in fact comparable. Investment per researcher is comparable, research output per researcher is comparable. We simply do not have as many numbers of researchers on the work bench as China. Lower research intensity is of course a concern, but it is growing slower and that is even a greater concern. This needs to be corrected. Of course, the Prime Minister of India made the statement in the last Science Congress that this will be corrected during the 12th Plan. Unfortunately, I have been hearing similar sentiments since the 9th Plan period onwards. Every Prime Minister of India makes exactly the same statement, but nothing changes.

While the standard S&T indicators are good indicators of the level of S&T activity in the country and often reflect the capacity of the country to embark on a technology-led economic growth, this capacity does not automatically translate into economic development and may remain dormant or get frittered away. I am not aware of any specific indicator of the forward link between R&D expenditure and economic returns. An indirect indicator of this forward link is the industry contribution to the national GERD (see Table 1.2).

Table 1.2

Composition of the National GERD

	(Average from 1995 to 2005 as a percentage of GDP)		
	USA	China	India
GERD	2.6	1.0	0.8
GOVERD	0.3	0.3	0.6
BERD	1.9	0.6	0.2
OTHERS	0.4	0.1	0.0

Source: RAND Report 'China and India, 2025', 2011.

It is interesting to note that India is the only country in this group where the industry contribution to R&D is much lower than that by the government. Industry investments on R&D have always been poor in India. Prior to the economic liberalization of the 1990s, this was not unexpected. However, even after economic liberalization, industry investments on R&D have remained poor. The industry neither plays a proactive role in the commercialization of the research results nor invests in R&D, and the government incentives to promote industry investments in R&D have had very marginal impact. While China is doing somewhat better than India, the contrast with the US cannot be missed. The US economy during the 20th century offers a textbook example of technology-driven economic development. The first five decades were dominated by defence technologies. On the other hand, the last three decades are being ruled by the marketplace: 'Silicon Valley' and 'Route 128' indeed characterize the three decades. The contributions of the Indian diaspora can also not be missed. Innovation and entrepreneurship are the mantras of these three decades.

Innovation, defined as the creation of novel and more effective products, processes, services or even ideas that are acceptable to the markets, cuts across the entire population of the country and is not related to the academic achievements of the persons. One indicator of innovation today is the number of patents filed. The US leads the world in terms of the number of patents filed across the world. Till recently, both India and China were poor in patents filing and notorious for their poor implementation of IP rights. But one must concede that civilizations that have survived for thousands of years cannot but be innovative. India and China are no exceptions. China has moved forward and has created a formidable IP portfolio. The last WIPO report of 2011 shows that while China is moving forward, India is yet to catch up. We have taken some steps; we have aligned ourselves with the global IP regime. We have sensitized our R&D system to protect IP. But, while a lot has been said or written about innovation in India, very little has been done to nurture innovation.

The National Innovation Foundation lists thousands of grass-root innovations. They, however, lack institutional support for validation, infrastructure for manufacturing and marketing. The TEPP program of the Ministry of Science and Technology offers financial assistance to the innovators cutting across the entire population. The Patent Facilitating Cell of DST offers information and assistance in protecting IP. All the initiatives are sub-critical. China's patenting system definitely started off earlier than ours, and they are doing better than us. We do not know how China is promoting innovation, but they are now filing nearly 75 per cent of the number of patents filed by the US. The government of India has recently declared the decade 2010–20 as the decade of innovations. I have high hopes on the National Innovation Council to put in place some concrete initiatives.

Entrepreneurship is the other link in the technology-transfer chain. An entrepreneur is the one who converts an innovation into a wealth generation activity. The economic impacts of entrepreneurship in MIT (Route 128) and in Silicon Valley are well known. Both India and China embarked on aggressive entrepreneurship development programmes very early. Unfortunately, over a period of time, India has lagged behind. I have been personally involved with the activities of the National Science and Technology Entrepreneurship Board and in the establishment of the Technology Development Board, both under the Department of Science and Technology. The role of NSTEDB in creating and nurturing a culture of entrepreneurship among young engineering students over the last few decades is indeed commendable. The Board has also established a chain of Technology Business Incubators (TBIs) in academic institutions to identify and nurture young entrepreneurs. It is worthwhile to note that while India and China embarked on the TBI programme almost at the same time, China has forged ahead both in the number and scale of operation of TBIs (see Table 1.3).

Table 1.3

Number and Scale of Operation of TBIs in India and China

	China	India
• First TBI started in	1987	2000
(Pilot Projects during 1987–90 in both countries with UNF S&T)		
• Number of TBIs (2008)	670	120
• Number of graduated tenants	31764	1150

Source: Author.

India has important lessons to learn from countries like Israel, Taiwan and, of course, China in promoting innovation and entrepreneurship.

As was pointed out earlier, India's GDP is lower than China's, and the contribution of technology to GDP is also definitely lower than China's. So, if we want to catch up with China in GDP, we have to increase production, manufacturing and marketing. There are a whole range of issues. Where are we investing? Where is the money coming from? Whereas in the United States, out of 2.6 for the GERD/GDP, only 0.3 comes from the government and 1.9 comes from the industry, in China, out of 1.0, 0.3 comes from government and 0.6 comes from business. In India, in contrast, out of 0.8, only 0.2 comes from non-government sources. So, clearly, Chinese industry is having better forward and better backward links to the industry. What they are doing is they are making high technology industries take the research output from the R&D system and they contribute back into the R&D system with much more ease than in India. This is one of the major drawbacks in our case, which has been talked about for the last 30 years. We are not discovering this now as it were: we knew it but we have not been able to make major corrections to it. An interesting exception is the increasing R&D investment by biotech companies in India (see Figure 1.1).

Figure 1.1

Average R&D Expenditure per Firm in India's Pharmaceutical Industry, 1992–2008

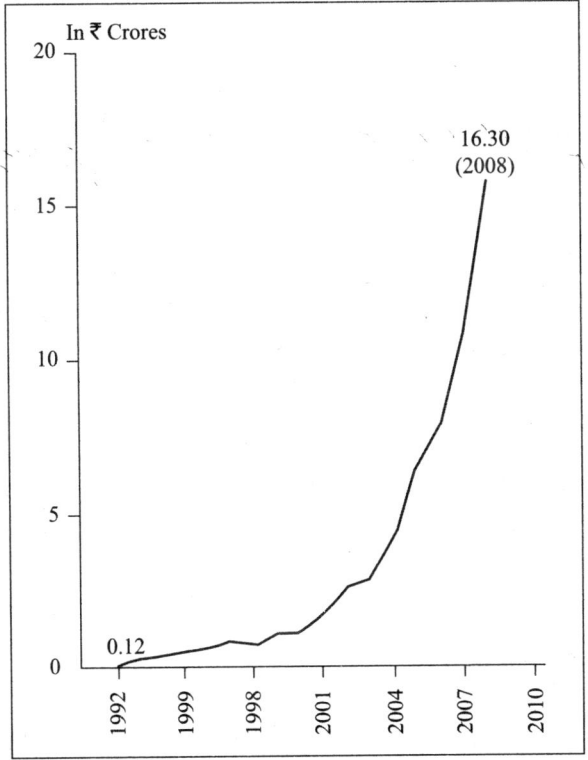

Source: UNESCO Science Report 2010: The Current Status of Science around the World.

This is the picture of R&D expenditure per firm in India's pharmaceutical industry; after 2000, it is just shooting up. I certainly believe that the early successes of some of the government initiatives in nurturing new biotech companies are responsible for this change in the scenario. But, this has not happened in the other sectors. The question to ask now is: why have not we expanded/replicated such programmes? We have not yet learnt to repeat success.

And, of course, the remnants of colonial rule in many of our government policies continue to act as a hindrance to our technology development and commercialization efforts. It sometimes takes years to approve a technology project. In an environment of fast-changing technologies this is not acceptable. Of course, it takes years to even close it if found unfeasible. Procurement policies are totally skewed, and the safety net for genuine technology failures does not exist! Ashok Jhunjunwala of IITM tells us how the wireless on Local Loop technology developed in India had to be sold through Brazil to India because China outbid the Indian company purely on the disqualification clause that they do not have the requisite experience of supplying 50 thousand lines. A new company does not start with supplying 50 thousand lines. It does not make sense, and there is no long-term strategy by government on industries in non-strategic sectors. This is something which we have to worry about.

Last but not the least, every link in the Mind to Marketplace technology-transfer chain is human-centric. The availability of skilled human resources is therefore a measure of a country's ability to create wealth and is as important as resources and capital inputs for any country with technological ambitions. While human resource is a truly renewable resource, it is education at all levels that converts a mouth to feed into a skilled pair of hands or an innovative brain. The human resource pipeline is also a pipeline that cannot be turned on and off at will as per requirements. Today's student is tomorrow's teacher who trains more students.

In general, China is presently better educated than India. India has an unacceptably large fraction of its youngsters outside the educational system at all levels. Even after 15 years of formal education, an unacceptably large fraction of our students turn out to be unemployable.

China not only has a much larger number of scientists and engineers per million population than India at present, it also trains more scientists and engineers annually. If we have to really draw the full benefit of the demographic advantage that we are likely to have over China in the 2020s and beyond, by increasing the number of youngsters in the productive age group and declining the dependency ratio, we need to revamp our educational system; and that too NOW.

2

Organization and Structure of Science and Technology in China

V.P. Kharbanda

China will expand its S&T personnel contingent. People who are engaged in RD activities will be raised from 1.05 million person-year in 2008 to 3.80 million person-year in 2020, and R&D personnel will be raised from 1.05 million person-year in 2008 to 2 million person-year in 2020.

The increase in the facilities to impart higher education in science and technology has helped in increasing the number of scientific and technical personnel enormously.

Introduction

Over the years, the growth of modern scientific institutions in China and the consequent formation of the Chinese scientific community have had a rather complex historical journey in the course of the six decades since 1949. Much of this complexity is centred around the relations between the state and the scientific organizations. Perhaps more than in any other society, the political context has severely conditioned both the goal and the direction of scientific research and the technological developments in China. Particularly, the commanding role of politics made some critical 'experiments'—the well-known Great Leap (GL) forward and the Cultural Revolution (CR)—in directing and shaping scientific and technological institutions to serve the social and economic ends set by it. In doing so, the organization of scientific and technological infrastructure had been subjected to many ups and downs in its evolution during the last six decades and is well documented in the literature. After the death of Mao, with the moderate leadership coming into power in 1978, and particularly after 1985, once again there was a shift to open policies and liberalization in the economy, calling for changes in Science and Technology (S&T) policies too to create the necessary conditions in S&T organizations to achieve desired social and economic goals, and the strategic goals of development.

This chapter attempts to explain the S&T organizations and infrastructure that have been re-shaped with the introduction of S&T policy reforms since 1985; and with the contemporary shifts to market mechanisms to meet the challenges posed by globalization and liberalization, how the S&T organizations have been allowed to evolve in the direction of independent innovation systems for the generation of proprietary innovations to meet the necessary conditions of socioeconomic development and compete in the international markets; and how the S&T policy changes in the new entrepreneurial environment of knowledge generation and commercialization are affecting the Chinese scientific institutions.

S&T Infrastructure and Its Internal Organization: Post-Liberalization Period (1985)

The conference on 'Science and Technological Work' held in March 1985 in Beijing led to new thinking and intentions on the part of the Chinese government to further unleash the forces of liberalization and technological modernization in scientific institutions. It was pointed out that science must serve the production system. Zhao Ziyang, in his speech delivered in the conference, stated that "economic construction must rely on science and technology while scientific and technological work must be oriented to economic construction".[1] He pointed out that the ineffectiveness of the research system, lack of transfer of technologies from research institutions to production units and acute shortage of scientific technical and managerial personnel were major obstacles to the achievement of these objectives. Several solutions were proposed to solve these problems, including the opening of technological markets, establishing technical contracts between research units and enterprises and proper utilization of scientific and technical personnel by improving their working and living conditions.[2] The use of economic levers and market regulations in the management of S&T was strongly advocated in order to enable scientific institutions to develop an internal impetus. The third session of the Sixth National People's Congress in April 1985 endorsed the proposals concerning S&T reforms.

After the government issued the policy document on 'Decision on S&T System Reform' in March 1985, the three main government sectors, namely, the scientific agencies such as Chinese Academy of Sciences (CAS), the universities and colleges, and the industrial enterprises constituted China's main strength in R&D by the early 1990s. In so far as science is concerned, the CAS took the lead in the all-round science reforms in China. Before we look into the main reforms which influenced the S&T research structure, let us briefly explore the decision-making structure in science as it prevailed in the early 1990s and thereafter.

S&T Decision-Making System

A schematic diagram of the science and technology organizations in China and their administrative structures is given in Figure 2.1. The top decision-making body responsible for S&T policies is the National People's Congress (NPC). Important acts or regulations concerning S&T development are required to be reviewed and ratified by the NPC. The State Council, as the top administrative organ of government, formulates administrative measures, issues decisions and orders, and monitors their implementation, drafts legislative bills for submission to the NPC or its Standing Committee and prepares the economic plan and the state budget for deliberation and approval by the NPC. It is the functional centre of state power and clearinghouse for government initiatives at all levels. It is responsible for carrying out the principles and policies of the Communist Party of China as well as the regulations and laws adopted by the NPC, and

Figure 2.1

A Schematic Diagram of the Science and Technology Organizations in China

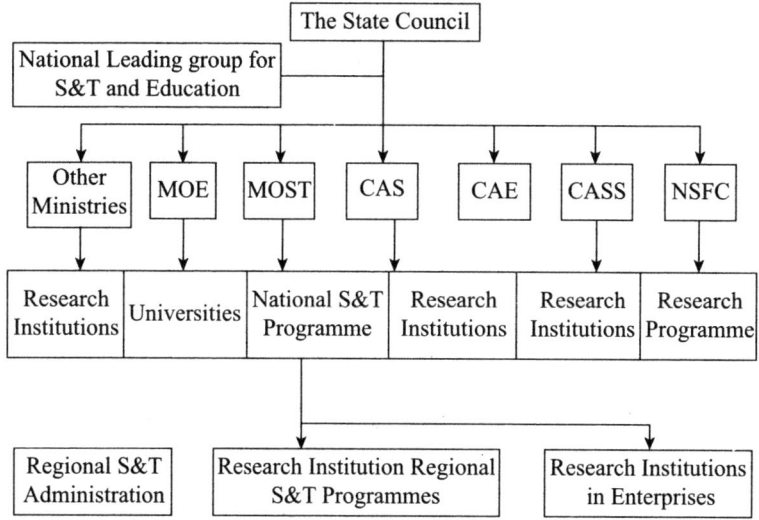

Source: Based on Mu Rongping, "Development of Science and Technology Policy in China", Mimeo. 13p., 2004.

Notes: MOE: Ministry of Education; MOST: Ministry of Science and Technology; CAS: Chinese Academy of Sciences; CAE: Chinese Academy of Engineering; CASS: Chinese Academy of Social Sciences; NSFC: National Science Foundation of China. (www.nistep. go.jp/IC/ic040913/pdf, accessed on 27 September 2010).

dealing with such affairs as China's internal politics, diplomacy, national defence, finance, economy, culture and education.

The State Council's 'Leading Group on Science, Technology and Education' was established in June 1998; it replaced the 'Leading Group for Science and Technology' established in 1983. The leading group, a special-purpose task force formed by the State Council to address problems that cut across administrative boundaries, is China's highest policy-making organ for science, technology and education. The leading group is mainly responsible for coordinating China's efforts to promote science and technology education, including research and drafting of national strategies and key policies, and for overseeing and implementation of national-level projects and campaigns and coordination among key PRC government agencies responsible for science and technology education. In addition, the group is responsible for the Medium and Long-term

National Plan for Education Reform and Development, which lays out fundamental directions, principles and key areas for China's education reform and development. The members of the leading group are drawn from senior levels—in many cases the minister level—from the ministries of Education, Science and Technology, Finance, and Agriculture, the Commission of Technology, Science, and Industry for National Defense (COSTIND), Chinese Academies of Science and Engineering and the National Natural Science Foundation (NNSF).

The S&T Structure

China's scientific research system is comprised of the Chinese Academy of Sciences (CAS), institutions of higher learning (including universities and colleges), industrial enterprises under various ministries, national defence research institutes, and provincial and municipal local-level scientific research institutes. Distinction was made between those institutes that were to concentrate on basic research, particularly in the CAS institutions and in the university sector and those which were mainly oriented towards applied research. There was a break with the Cultural Revolution policy where the whole scientific community was engaged in applied research, which was to be mainly concentrated in institutes under the ministries, and under provincial and sub-provincial levels. The over 160 national scientific and academic organizations affiliated to the China Association for Science and Technology, as well as its branches in various large and medium-sized cities, are also an important part of scientific and technological research.

The top administrative organization responsible for studying and formulation of important national S&T policies is the Ministry of Science and Technology. It is also responsible for providing guidance to various state commissions and ministries, the CAS, academies in agriculture and medical sciences, patent offices and S&T activities which take place at the provincial and municipal levels. It is responsible for managing and coordinating nationwide S&T activities. Its main functions include: to investigate and formulate national overall S&T strategies, as well as formulate guidelines, policies and regulations that foster S&T-based economic and social development, study and identify major layouts and priorities for S&T development, promote creation of national S&T

innovation system and enhance national S&T innovation capability; develop national medium- and long-term plans and annual plans for the civilian S&T development, work out policies and measures for strengthening basic research and new/high-technology development.

The Chinese Academy of Sciences

The Chinese Academy of Sciences (CAS) is China's highest academic institution as well as a comprehensive research centre in natural sciences. It is the highest advisory body in China on science and technology issues. It offers advice on the overall national science and technology development programmes and on decision-making related to major national science and technology issues; it carries out investigation studies on major science and technology issues faced in the national economic and social development. It offers suggestions on the formation of mid-to-long term development strategy and goals of different scientific disciplines and gives guidance and evaluation to the academic work in major research fields and key institutions.

The Scientific Council of the CAS is the highest authoritative organ in this system, composed of the members of CAS known as the 'academic department members' (*xuebu weiyuan*), introduced in 1955, who are nominated and elected from the most outstanding scientists from universities, government research institutes, as well as institutes of the CAS itself representing the cream of the Chinese scientific community. Membership is granted on the criteria of outstanding academic achievements, leadership in promoting their discipline, and 'patriotism' (understood as political reliability and loyalty to the cause of the people).[3] Selection of academic members, *xuebu weiyuan*, of the CAS departments is the top most lifelong honour with elite designation. New members are added every two years with a quota of no more than 60 for each election. In 1994, the CAS revised its by-laws and changed the title from academic members, *xuebu weiyuan*, to academician, *yuan shi*. The revival of the system of academicians, *yuan shi*, in the CAS thus was a major step to reward and recognize excellence.

There was no selection for academic members since 1957. As a result their number had reduced to 117 (from 254) due to attrition. In 1980, after 22 years, 283 new members were elected as academicians. In contrast to the 1955 and 1957 elections, in which the Party was deeply involved, the 1980 election was a merit-based selection by peers, with considerably less involvement of the Party and the government (Cong and Suttmeier 1999: 529).[4] In 1980, although the Party and the government tried to change the outcome of the election, the CAS members resisted their interference and the Party finally had to give up. Thus, there was a clear shift in the State policy to emphasize scienticism in contrast to the proletariat ideology pursued during Great Leap forward and the Cultural Revolution. As Wang (1991) also put it "by emphasizing the importance of scientists, researchers and technicians and by rewarding inventive work and talent rather than politics, the Chinese government hoped to raise the consciousness of the society as a whole".[5]

As the data on the CAS academicians shows (see Table 2.1), since 1991 there have been continuous efforts to institute, again, a reward and incentive system in the S&T system. This is indicated by the fact that institution of these norms of professionalization, through awards and honours, has resulted in a corresponding increase in the research publications and patents output at domestic and international levels.[6] Up to 2009, this honour had been given to 1143 CAS academicians.

The Presidium of CAS, chaired by the President of the CAS, is its Executive Committee. The six academic divisions of the CAS are: (a) Division of Mathematics and Physics, (b) Division of Chemistry, (c) Division of Biology, (d) Division of Earth Sciences, (e) Division of Technological Sciences, and (f) Division of Information Technical Sciences. The Academy, by 2010, had 97 research institutions reduced from the peak 124 in 1995 by mergers and closures due to the implementation of the policy of consolidation and merger of research institutions (Table 2.1), and it also had over 200 high-tech enterprises. In addition, it has over 20 Academy-affiliated supporting units such as one university and two colleges, five documentation and information centres, two printing workshops, five R&D centres for scientific instruments, etc. These institutes and units are located across the country. The CAS has invested in or created over 430 science and technology-based enterprises in eleven industries including eight companies listed on stock exchanges. There are also 12 CAS provincial branches at local level.

Table 2.1

Elected Academicians in CAS, 1955–2009

Department	1955	1957	1980	1991	1993	1995	1997	1999	2001	2003	2005	2007	2009	Total
Mathematics and Physics	30	6	51	38	10	10	9	10	10	10	8	6	6	204
Chemistry	22	2	51	35	10	9	10	8	10	10	9	6	8	190
Biological Sciences	60	5	53	34	11	12	12	11	12	11	12	7	5	245
Earth Sciences	24	3	64	35	10	10	10	10	9	10	7	4	5	201
Technological Sciences	36	2	64	68	18	18	17	16	15	17	9	5	7	292
Information Technological Sciences	–	–	–	–	–	–	–	–	–	–	6	1	4	11
Total	172	18	283	210	59	59	58	55	56	58	51	31	35	1143

Sources: Cong Cao, 'The Chinese Academy of Sciences: The Selection of Scientists into Elite Group', *Minerva* 36, no. 4 (1998): 323–46, 326; *Statistical Yearbook of Chinese Academy of Sciences-2000*, Beijing; *Kexue Chubanshe*, p. 312 (in Chinese); 'New CAS members elected', *Bulletin of the Chinese Academy of Sciences 2001* 15, no. 4 (2003): 214–15, http://English.people.com.cn/200311/24 (accessed on 11 October 2004).

Reforms at the Chinese Academy of Sciences

Under the guidelines set by the government in March 1985, S&T activities should serve economic development and economic construction, and the CAS, in its restructuring, introduced the following purpose in its various mechanisms.

Fund Allocation System

Till 1980, for funding research activities, classification management was followed. Since 1981, the CAS gradually decreased the weight of appropriations, and stressed that important research items and priority key projects should get more funding. Since the reforms in 1985, 35 per cent of its total funding was allocated based on the principle of preferential treatment to priorities. To implement this, comprehensive surveys would be made at various institutes and the proportion of three kinds of studies—basic studies, social- and public interest-oriented studies, and technical development—would be worked out. Over the years, funding for technical development has been decreased so that the enterprises become independent of government support. These were supposed to seek funding from the market and gradually become independent.

Revitalizing Mobility in Research Institutions and National Key Laboratories

Prior to 1985, the CAS was plagued with low mobility of personnel, which consequently caused rapid ageing of the staff, 'fossilization' of knowledge, and lack of interaction with universities and industry. This resulted in institutional compartmentalization, overlapping of research and low utilization of research facilities. In 1984, to break up the so-called 'institute ownership', the CAS established a number of new research facilities based on the principles of 'open, mobile, collaborating and serving the whole country'. The CAS opened 2 research institutes and 17 laboratories to scientists outside the Academy in 1985, which was increased to 65 laboratories and 8 working stations by the early 1990s. A number of institutes with modern facilities also opened their doors to researchers from the entire country.[7] These laboratories have not only consolidated China's research achievements in some disciplines, but have pioneered research in some newly emerging disciplines. They form a base for training and grouping of talented individuals and academic experts belonging to the young generation. A number of institutes with modern facilities opened their doors to researchers from the entire country.[8]

Since 1984, about 155 National Key Laboratories (NKLs) have been set up based in CAS and also in the universities with the support from MOST. These labs often specialize in particular areas of academic interest. These include: Ocean engineering, Chemistry, Medicine, Physics, Mathematics, Materials Science and Structural Engineering. To strengthen the development and management of NKLs, specific regulations were issued by the MOST and the Ministry of Finance jointly to revise the old by-laws, and the methods of NKLs operation and management were published on 29 August 2008. These included: (*a*) establishing earmarks for open operation, scientific instruments upgrading and proprietary innovations at the NKLs; (*b*) defining the procedures, requirements and methodology for screening research topics, selecting visiting scholars, upgrading scientific instruments and proprietary development in line with the scope supported by the earmarks; (*c*) setting up the principle and corresponding measures for the construction, operation and development of NKLs, namely, steady support, dynamic readjustment and evaluation on a regular basis; (*d*) strengthening daily management, especially annual plan and summary, and random check, and establish a pre-warning mechanism for the management; (*e*) further solidifying terms of reference of the academic committee, allowing it to play a proper role; (*f*) beefing up the responsibility of sponsoring agencies, stipulating that the sponsoring agency shall give priority support to the development of national key labs. The new management by-laws emphasize the importance of equipment sharing, data sharing, popular science diffusion, findings spin-off and collaboration with industry.[9]

Contract Management

From 1985, the contract system was introduced with collaborative studies on important S&T issues relating to the technological development of the national economy. In this system responsibility is defined between the contracting parties to ensure the completion of the tasks undertaken. The CAS took 47 major projects from the 'State S&T Priority Projects Plan' during the period of the Seventh Plan (1986–90) by using the contract system.

Establishment of Spin-Off Companies

To solve the issue of low value, poor record of enterprises in absorbing new technology and introducing new research in the market, CAS has created a number of hi-tech spin-off companies since 1984, mainly in the areas of IT, laser technology, pharmaceuticals, energy, new materials, electro-mechanical integration and biomedicine. Till the year 2000, over 400 spin-off companies under CAS were formed. With S&T as back-up and the market as the orientation, an integrated entity of a research, development, production and marketing has been formed. Some of the spin-off companies such as Legend Group, Shanghai Nicera Sensor Corporation, Chengdu Di'ao Pharmaceutical Corporation, Changchun Heat Shrinkable Materials Co. Ltd., Dalian Kaifer High-Tech Development Corporation, Shanghai Synica Chemicals Corporation, Shenyand Jinchangpu Company, and Huajian Electronic Co. Ltd. have made remarkable achievements. Many of the electronic and computer companies were established in affiliation with the CAS and institutes of higher education. These companies are concentrated in the street known as 'Electronics Street' (presently known as the 'zhonguangcun science and technology park' in Beijing. The other well-known companies which are in operation are: the Great Wall Consortium, the Beijing Stone Computer Company and Syntone Company. Most of these companies have been very successful. For example, Lenovo, founded in 1984 as Legend Group Ltd., is a spin-off of the Chinese Academy of Sciences which started with a seed capital of US$25,000 and a group of eleven scientists led by Mr. Liu Chuanzhi.[10] Lenovo brand PCs have been the best seller in China for seven consecutive years. By 2009, CAS had invested in or created over 430 science- and technology-based enterprises in eleven industries including eight companies listed on stock exchanges.

One Academy with Two Running Systems

In the course of the reforms, since 1988, two totally different systems have come to stay in the CAS as "one Academy two operational Systems". One is the basic scientific research system with activities stemming from the research itself. This segment also addresses the strategic and fundamental issues relating to social and economic development. The other is the technical development system which is concerned with the development of technological products in engineering, new products and marketing. Since 1988, separate assessment standards, running mechanisms and management have been introduced to assess the activities in the two mechanisms of the CAS. Practice over the last few years has shown the viability of duality in the activities of the Academy.

Basic research has always been one of the major tasks of CAS. Over the years, the CAS has set up various branches of natural sciences within itself, including mathematics, physics, chemistry, mechanics, astronomy, space science, life sciences, earth sciences and environmental sciences, and has formed a contingent of about 10,000 people for basic research. The CAS has established 117 laboratories, which are open to competition by scientists from both home and abroad. Since the 'Sixth Five-Year' Plan (1981–85), with the approval of the State Council, the CAS had constructed some large research facilities, such as the electron positron collider in Beijing, heavy ion facility in Lanzhou, synchrotron radiation facility in Hefei, Tokmak and laser fusion devices, the 2.16 m optical telescope, and the Beijing solar magnetic field telescope, and the Shaanxi long-wave time service station.

The present priority research areas of the Academy are: large-scale integrated circuits, new materials, bioengineering, laser, information sciences, oceanic survey and development of marine sources, computers, software and superconductive technology. The priority of research in basic biology includes molecular biology, cell biology, neuroscience, developmental biology, genetics, genomics, proteonics, psychology, etc. Important accomplishments have been made by CAS scientists such as the first total synthesis of crystalline bovine insulin and the total synthesis of yeast alanyl-transfer ribonucleic acid in the world, the physical mapping of rice genome and sequencing of No. 4 chromosome DAN, the participation in the International Human Genome Project as well as taking the lead position in the completion of the working draft of 1 per cent human genome sequence.

In March 2008, the Chinese Ministry of Finance and the Chinese Academy of Sciences jointly announced that they would finance the proprietary development of eight key research facilities which are: equipment for producing deep ultraviolet all-solid-state laser source, a pulse wind-tunnel that is able to imitate supersonic flying condition, a comprehensive experimental system for extreme conditions, a mobile seafloor seismological observation array, a superconducting imaging spectrometer, a digital VLBI base band converter, a nanometer-based synchrotron radiation imaging equipment and an intermediate energy heavy ion micro irradiation unit. Experts believe that the planned development will raise China's research capability on the one hand and gather more experience for developing proprietary research facilities on the other.[11]

Promotion of Excellence

In 1981, the revival of the system of academicians (*yuan shi*) in the CAS was a major step to reward and recognize excellence. In 1994 the CAS also revised its by-laws and changed the title from academic members, *xuebu weiyuan* to Academician *yuan shi*. Since 1995 election, these academic members are recognized as 'Academicians of the CAS'. These scientists and engineers constitute the new scientific elite. This system served to provide professional leadership in science and technology and advice to national decision makers on socio-economic problems. The system as a whole has revived the support to values of scientific elitism prevalent in the early fifties. These academicians have the responsibilities of:

a) recruiting members through peer review from among best scientists and engineers as the basis of professional merit;

b) maintaining a close relationship with political elites;
c) advising on national policy matters involving science and technology; and
d) providing academic leadership to the scientific community at large.

The system of academicians thus recognizes and utilizes the scientific community not only to use them in socio-economic development but also to promote excellence within the scientific community. The reconstitution of institution of scientific leadership was essential in building up of scientific community in specific disciplines.

During four recent elections, elite scientists have actually used the autonomy given by the Party to elect those they considered to be qualified, and the elections are now completely free of Party control. In practice, the number of papers a scientist has published, the reputation of the journals in which publications appeared and the number of citations to a scientist's work have been used as important measurements of the quality and impact of the work of a candidate. Awards a candidate has received at the national, provincial or ministerial levels also serve as evidence of his or her achievements or contributions. In addition, invitations to speak at international conferences are also used as an important criterion.[12] Thus through this academic system, academicians have come to pay more attention to professional integrity and attitudes towards research in the selection of candidates. The elite body is thus conveying the message that Chinese scientists are expected to observe norms and rules for scientific conduct that are adopted by the international scientific community.

National Knowledge Innovation Programme

The most dramatic case of the 1995 institutional reform policy is the 'Knowledge Innovation Programme' (KIP)[13] of the CAS. The Leading Group decided that the CAS take the lead in launching a pilot project of the National Knowledge Innovation (NKI) Programme with the objective to build up a National Innovation System and make China an innovation oriented country. In order to turn the CAS into a scientific research base of international advanced level, a base for the training of high quality scientific talents and a base for the promotion of the development of high-tech industries in China, in early August 1998 the CAS decided to undertake 12 assignments of reorganization as the tasks to be done in the initial phase of the pilot project of the NKI programme. Pilot reform initiated in 1998 was selectively launched in institutes, such as the Dalian Institute of Chemical-Physics, Institute of Theoretical Physics, Nanjing Institute of Geology and Paleontology, Changchun Institute of Optics, Fine Mechanics & Physics, etc. Efforts were made in pushing the institutes to establish mechanisms that could link knowledge innovation with the industrialization of high technology. Up to now, 10 institutes have been identified as the first group to undertake the whole structural transformation.

In basic sciences, in areas such as mathematics and life sciences, efforts were made in transforming institutes into academic centres. Accordingly, it established the Academy of Mathematics and System Science, the Shanghai Academy of Life Sciences, Beijing Research Base for Physical Sciences, the Center for Theoretical Physics, the National Astronomical Observatory Centre and R&D centres for information science and technology in Beijing. Apart from this, efforts are being made to tighten the collaboration between institutes and also between institutes and enterprises: there is also the formation of three innovation bases in the field of natural resources and environment in Beijing, southwest and northwest of the country in accordance with their regional features and academic strength by integrating existing institutes. The success of the implementation phase II (2001–05) was shown by the fact that China's competitiveness in the world advanced from 30th in 2000 to 24th in 2004, with its overall S&T strength topping the developing countries.[14]

One of the important components in the Knowledge Innovation Programme is to speed up the commercialization of research results to promote high-tech industrialization. In this direction, the CAS, while transferring its mature technology to various domestic industries, has also created high-tech enterprises of its own as discussed above. The R&D work of the CAS in high-tech development involves the fields of information technology, advanced manufacturing, optics and fine mechanics, materials, energy, transportation, chemical engineering, space and remote sensing technology, etc. In order to strengthen the ability of the institutes in technology transfer and commercialization, the CAS also set up a number of 'Engineering Research Centers'. So far, the CAS has also established cooperative partnerships with over 3,000 enterprises and has jointly set up 'Technological Development Centers' with medium and large enterprises in an effort to vitalize unused resources from both sides and to strengthen the technical innovation capability of enterprises.

At the provincial level, the CAS has set up a 'Collaborative Fund between CAS and Provinces and Cities' and 'Collaborative Award between CAS and Provinces and Cities'. It sent some of its scientists and engineers to the local governments and enterprises to serve as deputy heads for science and technology and to build up research teams, construction of research sites and establishment of an innovation culture.

The goal of the CAS by 2010 was to build itself into a National Knowledge Innovation Base in natural science research and high-tech development, which possesses the ability to make innovations on a continuous basis and is capable of conducting research at the frontiers of science besides helping solve practical problems of economic construction and social development. It also aims to become a national bank of scientific knowledge, a think-tank of scientific thoughts and a pool of scientific talents of international level. Speaking on achievements of the KIP programme on 30 October 2009 at the 60th anniversary conference of CAS Lu Yongxiang, the President of CAS pointed out that since the implementation of the KIP Programme, the CAS has fostered nearly 1,000 new-generation S&T leaders and top rank talents and formed a high-profile scientific innovation team, among whom there are 600 chief scientists or principal leaders for major national scientific tasks, nearly 700 winners of National Science Fund for Distinguished Young Scholars, 53 innovation groups evaluated by and financially supported by the NNSFC and 900 scientists who have assumed key posts in major international science organizations.

The Institutions of Higher Learning

The second important sector of government in R&D is the Chinese universities and colleges. Even though this sector did not play a significant part in the national research system in the past, the importance of this sector has grown suddenly from the beginning of the 1980s when the postgraduate courses began after the CR era.

Zhao Ziyang, speaking at the National Science and Technology Conference held in March 1985, stated: "The biggest obstacle still lies in talented personnel—we lack scientific, technical and managerial personnel." In this direction, reforms in the management of scientific personnel have been constantly emphasized to create a large pool of talented scientific and technological personnel who can effectively utilize their specialized knowledge through entrepreneurship. China has been making considerable efforts in this direction in the last three decades. In the legacy of CR, this problem has been accentuated by: (*a*) limited higher education facilities (only 50 per cent of students coming out of schools are able to enroll in college for graduate education); (*b*) large-scale brain drain to Western countries (of the 134,000 Chinese students studying abroad in 2006, only 42,000 returned), see Table 2.2; and (*c*) ageing.

Table 2.2

Number of Students Studying Abroad

Year	Studying Abroad	Returned to Mainland	Percentage of Returning Students
1957	529	347	65.59
1978	860	248	28.8
1985	4888	1424	29.13
1990	2950	1593	54
1995	20381	5750	28.2
1996	20905	6570	31.42
1997	22410	7130	31.8
1998	17622	7379	41.9
1999	23749	7748	32.6
2001	84000	12000	14.3
2002	125000	18000	14.4
2003	117000	20000	17.0
2004	115000	25000	21.7
2005	119000	35000	29.4
2006	134000	42000	31.34

Source: Various issues of *China Science and Technology Newsletter*, *China Daily* and *Beijing Review*.

Because of continued reforms, the number of Institutions of higher learning has increased considerably, in fact, more than tripled from 598 in 1978 to 2,236 by the end of 2008 (Table 2.3), along with increase in total enrolment at various levels. In 1982, the enrolment to post graduation in science grew to 21,184 (per year), of which 20,910 were for master's degrees and 374 for doctorates.[15] Over the years facilities for postgraduate education have increased tremendously. In 2008, postgraduate enrolment increased 500 times than that in 1980. In the

Table 2.3

Enrolment in Institutions of Higher Education

Year	No. of Institutions	Graduate Enrolment		Post Graduate Enrolment	
		During the year	Total	During the Year	Total
1949	117	NA	117	NA	NA
1964	277	750000	NA	1456	4546
1975	NA	NA	501000	NA	NA
1978	598	230000	856000	10500	11000
1980	633	280000	1134000	8000	21600
1985	1016	619000	1806500	47000	87000
1990	1075	604000	2063000	30000	93000
1995	1054	926000	2906000	51100	146000
2000	1041	2206000	5561000	129000	301000
2001	NA	2680000	6872000	165000	423000
2002	2000	3210000	9030000	203000	500000
2005	1778	NA	15000000	400000	1100000
2008	2263	6076612	20210249	386658	1046429

Sources: Various issues of *China Daily, China Science and Technology Newsletter* and *Beijing Review;* Education in China—20 years of Chinese communism, 1969, p. 30 (mimeo); Communiqué of the State Statistical Bureau of the PRC on fulfillment of China's 1981 National Economic Plan, 29 April 1982; China's Higher Education for the Next Century, Beijing; *xin sheng chuban she chu* (in Chinese), 1995, p. 19; http://202.84.17.71/english/china - abc/educal.htm downloaded on 28 March 2002; http://www.china.org.cn/c-internet/KJ/htm/20-J.htm downloaded on 28 March 2002; http://www.edu.cn/ downloaded on 5 April 2002; China Statistical Yearbook on Science and Technology—1999. Beijing; *zhongguo tongji shubanshe,* p. 243 (In Chinese); China Facts and Figures 2003:1–120; State Council Information Office 2003: 1–199; Beijing Review 2010, 53 (22).

same year China surpassed the US in the production of PhD degree holders too (Yao Bin, 2010). The data clearly shows that enrolment in graduate and postgraduate education in the year 2008 doubled over the year 2002. Corresponding with the merging of many public universities, there has also been the rapid expansion of the private sector since 1999. As of 2006, private universities accounted for around 6 per cent of student enrolments, or about 1.3 million of the 20 million students enrolled in formal higher education (Wikipedia, accessed 29 September 2010).

Academic excellence is also being stressed above political purity and the university staff is expected to be involved in basic as well as applied research apart from teaching. It was also expected that the multidisciplinary character of universities will enable them to organize coordinated research work in frontier areas and newly emerging technologies. It emphasized that research, no matter how basic in character, cannot be research for its own sake: it must be integrated with production.

The system of having 'key universities' and 'ordinary universities' was also restored. In 1980, out of 633 universities, there were 89 key universities, which admitted students of high excellence. The key universities have advantage over ordinary universities of having better facilities, equipment and teaching staff[16] and are under the control of the Centre, while ordinary universities are under control of provinces and municipalities, which are less selective in their admission policies.[17] By the year 2000, the Chinese government instituted a two-tier system of universities, namely, universities managed by the central government and universities managed by provincial governments. With this the majority of universities are managed by provincial governments. A small number of universities (about 75 now) which were highly specialized are managed by the Chinese Ministry of Education (or by a few other ministries). Presently, China has 107 key universities.

In 1995, the Ministry of Education of the People's Republic of China also initiated 'Project 211' for the establishment of 100 universities and colleges with the aim to bring them among the advanced universities in the world by the beginning of the 21st century. The project was incorporated as a national key development project during the ninth Five-Year Plan (1995–2000) for implementation. It primarily aimed to train high level professionals to meet scientific, technical and human resource standards, to offer advanced degree programmes and push forward the development of key disciplinary areas of science and technology.[18] It was an initiative funded by the State Development & Planning Commission of China to improve the quality of education, research, management, and institutional efficiency. Two hundred and eleven project schools have taken the responsibility of training four-fifths of doctoral students, two-thirds of graduate students, half of students from abroad and one-third of undergraduates. They offer 85 per cent of the state's key subjects, hold 96 per cent of the state's key laboratories, and utilize 70 per cent of scientific research funding. During the first phase of the project from 1996 to 2000, approximately US$2.2 billion was allocated to it. In May 1998, 'Project 985' was also launched by President Jiang Zemin, who called for founding world class universities in the 21st century. In the initial phase, it included nine universities, and by the year 2004 this number increased to 40.[19]

The increase in the facilities to impart higher education in science and technology has helped in increasing the number of scientific and technical personnel enormously. The S&T manpower increased from 0.59 million in 1978 to 10.809 million in 1990 and to 21.651 million in the year 2000, showing a doubling effect over the two decades. However, the data according to China Science and Technology Data Book, 2007, issued by the MOST, shows that the total S&T manpower increased from about 3.2 million in 2001 to 4.96 million in 2008 (Table 2.4). This indicates that the figures for S&T manpower before the year 2001 were inflated to include the technicians too.[20] There has also been considerable increase in R&D manpower also, which increased from

Table 2.4

Stock of Science and Technology Personnel in China (2001–08)

(10,000)

Year	2001	2002	2003	2004	2005	2006	2007	2008
S & T Manpower	314.1	322.2	328.4	348.1	381.5	413.2	454.4	496.2
R & D Manpower	95.7	103.5	109.5	115.3	166.5	150.2	173.6	196.5
Scientists and Engineers	74.3	81.1	86.2	92.6	111.9	122.4	142.3	159.2

Source: Various issues of *China Science and Technology Newsletter; China Statistical Yearbook, 2009,* National Bureau of Statistics of China, Beijing.

0.310 million in 1978 to 0.957 million in the year 2001 to 1.965 million in 2008. While the number of scientists engaged in R&D activities is small in absolute terms, the number of 'full time equivalent' solely engaged in actual R&D work is much smaller if such factors as time spent on activities unrelated to S&T (for example, administrative meetings), teaching, consultancy and other S&T activities and work that could be done by lower levels of staff are taken into consideration.

In order to tap potential and raise efficiency, a number of universities and colleges have begun to join their capacities with each other. For example, Jiangxi University and Jiangxi Polytechnical University merged to become Nanchang University; Sichuan University and Chengdu University merged to become Sichuan United University. In 1995, more than 200 universities and colleges had either become merged or united or were planning to do so.[21] In 1998, the Zhejiang University became the largest university in China by combining with it the province's medical and agriculture universities and the Hangzhou University.

Another measure to raise the quality of education was in the direction of raising remunerations for the faculty. In November 1999, the teachers' pay in the Tsinghua University was revised considerably. Low pay for the university teachers had made the profession less attractive and become the number one reason for the brain drain among the teaching staff. It is reported that each year 10 per cent of the country's 78 key universities' teachers leave for other institutions. The monthly pay of the university professors accounted for only 1 per cent of their counterparts in Hongkong. The reforms are being led by Peking University, Tsinghua University, Fudan University, Shanghai Communications University, Chinese University of Science and Technology, Zhejiang University etc.[22]

Like the CAS, among the national universities, Beijing University too has established a number of spin off companies, for example, the Bei Da New Technology Company (founded in 1988), the Bei Da Fang Zheng Co. and Bei Da Weimin Bioengineering Co.[23] Their product—the Chinese character laser computing system developed by Professor Wang Xuan—is considered to be a great breakthrough and is acclaimed as the second revolution in Chinese printing industry.[24] According to reports, per capita R&D expenditure by these university-run enterprises was about US$ 850 in 1995,[25] which is much higher than the average per capita expenditure on R&D in China. The success of these university-run enterprises is evident from the fact that some of them have been listed among the country's top 100 high-tech enterprises. These are Beijing University's Founder Group, Xi'an Jiaotong University's Kaiyuan Group, Qinghua University's Ziguang Group, and Fudan University's Fuhua Group.[26] These spin-off companies from the CAS and universities have thus established a new culture of 'scientist-entrepreneurs' with market orientation.

University Science Parks

In order to efficiently utilize the university research results into production, China is also establishing 'science parks' in the universities. The Chinese Ministry of Science and Technology and the Ministry of Education have recently released a 'National University Science Park' development plan for the twelfth five-year period (2011–15), in an attempt to sustain the development of university science parks in the new five-year period. According to the plan, China will have 200 university science parks under a three-tiered management system by 2015. By that time, university science parks at the national level will hit 100 in number with 10 million square metres of floor space. Professional service firms stationed at the parks will reach 1,000 in number, and businesses to be incubated 8,000 in number. The twelfth five-year period will see the graduation of some 5,000 businesses and 100,000 service firms, with 10,000 S&T findings transferred, and train 100,000 innovation and entrepreneurship talents. Meanwhile, university science parks will build 80 centres for students' tech start-up practice, and establish 3,000 students' tech start-ups. Main objective is to promote technology transfer and finding spin-offs, taking advantage of the universities' innovation strength, helping faculties and students create tech start-ups, improving services, and fostering innovation and entrepreneurship talents.[27]

The above measures to reform the education system have helped to increase the Chinese S&T strengths to a considerable extent. In 2008, China surpassed the US as the world's top producer of PhD holders. However, attending an academic forum, the Chinese Minister of Education Zhou Ji admitted that the expansion of educational enrolment programme has also given rise to various problems, including the low quality of education. On the 2010 list of the world's top 100 universities issued by the US News and World Report, the highest ranked Chinese university, the Peking University, is ranked at the 52nd place, followed by the Tsinghua University in the 56th place. According to Xu Zhihong, former president of the Peking University, Chinese universities still lag behind world-class universities, with most evident gaps in teaching techniques. Achieving the goal of building world class universities in China may take longer than two generations of time.[28] The major problem indicated is the far-reaching government interference in the administration of universities, as if these are government agencies, which decide on the results of the rating of universities, the appointment of faculty and staff, allocation of funds, designing of curriculum, how to use funds and how to evaluate universities.[29]

While various measures have been taken up to expand higher education enrolment and lure overseas Chinese students, the problem of leadership positions is still causing difficulties. The majority of scientists engaged in the forefront of scientific research at various research institutions, who were university graduates including a small number of postgraduates from the 1950s and 1960s, have retired, and that obviously has led to widespread 'talent fault'. The average age of Chinese researchers at the Chinese Academy of Sciences dropped by nearly 10 years between 1991 and 2003, as the older generation retired and younger researchers, many educated in the United States and other foreign countries, took their places. Because of the ageing scientific community, there are very few scientists with leadership qualities. A survey done in 2002 on biological community by the author indicates the scarcity of leadership positions in biological sciences in China. Out of 64 biological scientists surveyed, only one is at or above the age of 56 years and only nine were between 46–55 years. Out of the lot, only six (nine per cent) were holding doctorate degrees. However, most of the scientists (84.38 per cent) are below the age of 46, and about 46 per cent of them have doctoral degrees. In total, only about 56 per cent of the scientists hold PhD degrees, clearly indicating the scarcity of expert manpower—a legacy of the Cultural Revolution, which had a detrimental effect on higher education and academic activities. However, the drastic increase in young scientists having doctoral degrees after the revolution shows that enormous efforts are being made to fill this generation gap after that restoration of professional climate in 1978.[30] There was a problem at the entry level too. According to statistics from the Ministry of Education, currently only 10.5 per cent of young people between 18–24 can receive higher education, nearly 40 per cent lower than the average level in developed countries (University intake, 2000, p.2).[31]

To fill the gap quickly this problem of ageing in research institutes is being addressed through intensive training and education of the younger generation. In an attempt to strengthen higher education, a series of measures have been adopted, such as, the 'Young Scientists Fund', 'The Special Fund for Middle Aged and Young Scientists' and the 'State Outstanding Young Scientists Fund'. Sixty research bases from 30 higher institutions were supported by the Fund for Talent Training in Basic Science with a funding of 99.30 million Yuan. Actively accommodating the needs of rapid development of China's young scientific and technological teams, NSFC funded 3,336 Young Scientists Fund projects, with a funding of 617 million Yuan. NSFC's stress on talent fostering has accelerated the growth of team construction for China's basic research.

Recently, Chinese authorities released a detailed plan to implement 12 major talent projects[32] defined by the National Medium and Long Term Talents Development Planning (2010–20), aiming at the following objectives:

1) **Innovation Talents Programme**: Led by the Ministry of Science and Technology, by 2020 the programme is designed to establish 100 scientists' studios, produce 3,000 young and middle-aged tech innovators, finance 1,000 qualified entrepreneurial talents on an annual basis, establish 500 innovation teams in the priority areas and build 300 innovation role model training and demonstration centres.

2) **Young Talents Development Programme**: The programme is created by the CPC Central Committee Organization Department to nurture 2,000 top-notch young talents under the age of 35, screen out 1,200 top-notch students from renowned research universities on an annual basis, and select 2,000 outstanding high school and college graduates to further their training at renowned overseas universities.

3) **Capacity Building of Business Management Personnel**: The programme is sponsored by the state-owned Assets Supervision and Administration Commission to bring out, by 2020, 500 entrepreneurs with global vision, strategic thinking, innovative spirit and business ability, and nurture 10,000 high-calibre business management personnel who are good at strategic planning, capitalization, human resources management, accounting, and legal matters.

4) **High-Quality Educator Training Programme**: Initiated by the Ministry of Education, the programme is meant to train 20,000 school teachers, disciplinary leaders, and principals on an annual basis, and nurture educators, famous teachers, and disciplinary leaders for primary and secondary schools (including kindergarten), vocational schools, and universities.

5) **Famous Cultural Talents Programme**: Led by the CPC Publicity Department, the programme will created to fund 2,000 renowned specialists in the areas of philosophy, social sciences, journalism, publishing, radio and television, culture and arts, cultural heritage protection, cultural business management and cultural technologies, allowing them to be part of major researches, projects, performances, creative researches, exchange and shows, monographs publication among other activities.

6) **Health Talents Programme**: Initiated by the Ministry of Health, the programme is designed to foster by, 2020, high-calibre medical personnel and securing special funds for related earmarked researches. Efforts will be made to standardize residency training, and bring out 50,000 resident physicians for different disciplines. Meanwhile, some 300,000 general practitioners will be trained under the programme.

7) **High Calibre Overseas Talents Programme**: The programme is sponsored by the CPC Central Committee Organization Department to implement a range of related programmes or projects, including a 'Thousand Talents Program' at the central level, short and long term innovation projects, humanities and social sciences projects, a 'Thousand Young Talents Program', a 'Thousand Foreign Experts Program', and business pioneering projects among others. It plans to attract high calibre overseas talents to establish their own businesses in China in 5 to 10 years.

8) **Professional Knowledge Updating Project**: The project is a large knowledge-updating campaign established by the Ministry of Human Resources and Social Security to train 100 million senior specialists in 12 areas, including equipment manufacturing, information technology, biotechnology and new materials. Meanwhile, efforts will be made to establish a number of national further education centres for professionals and technical personnel.

9) **Highly Skilled Personnel Programme**: Created by the Ministry of Human Resources and Social Security, the programme will, by 2020, bring out 3.5 million new technicians and one million senior technicians, raising the total number of technicians and senior technicians in the country to 10 million. Meanwhile, it plans to build 1,200 training centres to bring out more highly skilled personnel.

10) **Modern Agriculture Talents Programme**: The programme is initiated by the Ministry of Agriculture to finance, by 2020, 300 high-calibre researchers in the area of agriculture, and to support 10,000 personnel who have made outstanding contributions to diffusing advanced agricultural techniques.

11) **Human Resources Support Programme for Remote, Poverty, Ethnic and Veteran Revolutionary Areas**: Sponsored by the CPC Central Committee Organization Department, the programme is designed to guide, by 2020, 100,000 outstanding teachers, doctors, scientists, technical personnel, social workers, and cultural workers to work in, or provide services to the remote, poverty-affected, ethnic and veteran revolutionary areas on an annual basis. Efforts will also be made to annually train 10,000 urgently needed talents for those areas.

12) **College Graduates Training at Grass-Roots Programme**: The programme is created by the CPC Central Committee Organization Department to work on a range of sub-programmes for college graduates to work as village officials. It will also create ad hoc positions for rural school teachers, provide free teacher education and training, free medical students training, prepare college graduates for working in the rural areas and student volunteers to provide services to the western region.

Promotion of Institutional Incentives and Reward System

In scientific organizations, awards and incentives are the essential norms as an inducement to research activity for professionalization. In this direction in 1995, the CAS instituted a system of incentives and awards for persons who had contributed significantly to knowledge or to the national economic development through their research work or scientific writing. After the demoralization suffered by scientists in the Cultural Revolution, the decade of the 1980s witnessed infusion of new confidence in the scientific community with the initiation of S&T reforms. In 1978 Hua Kuo-feng stated:

> Commendations and proper awards should be given to those units that have achieved marked success in adopting new techniques and develop new technologies and turning out new products as well as to those collectives and individuals who have made inventions.[33]

Since then, the national reward system has been gradually institutionalized, enhanced and diversified over the last three decades. In the mid 1980s, China established three annual awards: (*a*) The National Award for Natural Sciences, (*b*) the National Award for Technological Inventions, and (*c*) the National Award for Scientific and Technological Progress. The first prize for the three honours is of 90,000 Yuan (US$ 10,000) and the second prize is of 60,000 Yuan (US$ 7200). Four hundred awards are given each year. In 1994, the Ho Leung Ho Lee Prize, supported by the Hong Kong-based Ho Leung Ho Lee Foundation was initiated. The two awards Award of S&T Achievements (US$ 128,000) and the Award of S&T Progress (US$ 26,000) are both affiliated to this Prize. Up to 1999, 311 scientists had been honoured with these awards (Jia, 1999, p.2).[34]

The National Committee for Scientific and Technological Awards (NCSTA) was established in December 1999 with the approval of the State Council. The principal duties of the NCSTA are to control and direct awards for science and technology, organize the National Appraisal Committee for scientific and technological awards by engaging relevant specialists and scholars, pass resolutions on award winners, types and levels of awards, and policy-related suggestions for perfecting the presentation of wards.[35] During the same year, the State Supreme Scientific and Technological Award (US$ 730,000) was also approved by the State Council for making significant contribution to leading scientific fields, technological innovations and applications and high-tech development . Ten per cent of the amount will be given directly to the winner and the remainder for the scientific research project chosen by him. This award is presented by the President each year.

Research Publications

One of the most important activity of the members of scientific communities is to generate and disseminate knowledge in the form of publications and research papers. Publication of scientific journals and research papers has resumed tremendously after the Cultural Revolution, particularly since 1985, with the revival of the policies for professionalization and free academic environment. There has been continuous and rapid increase in the publication of research papers in China since then: in 1998 it surpassed the number of research papers published by Indian scientists (see Table 2.5).

Table 2.5

Numbers of Scientific Papers Published in India and China

Year	India	China
1980	14983	924
1982	12124	2592
1984	10600	2537
1986	10854	3678
1988	10208	5312
1990	10103	6509
1992	11160	7630
1994	11319	8226

(continued)

Table 2.5

(continued)

Year	India	China
1996	11177	10152
1998	12128	14610
2000	12127	22061
2007	30000	94800

Source: Based on Arunachalam 2002: 108; China S&T Statistics, 2007; Jonathan et al., 2009.

Note: Data as seen from Science Citation Index.

At present China also has a much larger number of scientific journals as compared to India. According to Ren, there were about 4,294 scientific journals[36] in China in 1999,[37] while India had almost half of that number i.e. 2,255 in the year 2000.[38] In the same year, *Science* reported 4,300 Chinese scientific journals.[39] This large number of scientific journals in China is attributed to the fact that every university, institute and professional society in China has at least one and frequently more publications. However, many of these journals seem to be of low international quality and, as a result, Chinese journals are lagging behind those in the rest of the world. This is shown by the fact that the number of Chinese scientific journals covered by the Institute of Scientific Information, Philadelphia (ISI) has been decreasing over the last 15 years.

However, as shown by the Science Citation Index (SCI) of the ISI, there has been considerable increase in the research papers published by Chinese scientists from 35,685 in 2001 to 94,800 in 2007 (Table 2.6). In 2003[40] itself, China had improved its international standing to 5th place.[41]

The Industrial R&D

The third R&D segment consists of the government sector comprising large and medium enterprises or industries, which are about 11,000 in number, employ about 700,000 scientific and technical personnel (Table 2.7). In 1998, these institutions numbered 10,926 after reaching

Table 2.6

Number of China Science and Technology Papers as Indexed by SCI, EI, ISTP (2001–07)

Year	2001	2002	2003	2004	2005	2006	2007
SCI	35685	40758	49788	67377	68226	71450	94800
EI	18578	23224	24997	33500	54362	65142	78200
ISTP	10263	13413	18567	20479	30786	35463	45331
Total	64526	77395	93352	111356	153374	172055	218331

Source: Various issues of *China Science and Technology Newsletter*, *China Science and Technology Statistics Data Book* (zhongguo kexue tongji shuju), 2007.

Table 2.7

Number of R&D Institutions by Sector

	1990	1993	1995	1996	1997	1998
Govt. Institutions	5084	5119	7721	7636	7558	7496
Higher Education	806	814	3431	3398	3343	3241
Enterprises	8116	10477	13107	12033	11142	10926

Source: S&T sector, www.edu.cn/ (downloaded on 7 December 2001).

a peak of 13,107 in 1995, which means that a number of inefficient industrial R&D institutes were closed down.[42] The R&D institutions sponsored by the collectives and by individuals although have undergone rapid growth in numbers in recent years, but seldom made any significant achievements.[43] These private research institutes are too small to conduct any meaningful research, because of scarcity of material and human resources; some of them have only one to two persons. In general, enterprise-sponsored research institutions mainly serve the technological progress of their own enterprises and are devoted to market oriented projects which yield quick benefits and are of short duration as well as easy to develop. In view of the weak R&D capabilities of the enterprises, the reform policies over the years have advocated the strengthening of linkages between S&T institutions and production enterprises, and secondly, the transforming of production oriented R&D institutes into S&T enterprises.

Research–Production Linkages

The reforms decision called for more effective research–production linkages by two ways:

First, by forcing R&D institutions into economic accountability and also by drastically cutting their budgets, thus making them susceptible to technological change. In this direction, Teaching-Research-Production (T-R-P) Associations were formed to help facilitate transfer of technology for new products and processes developed by the academic institutions to small and medium size or village and township enterprises. Since 1990, 74 National Engineering Research Centers (NERCs) have been established to help transfer of technologies from universities and R&D institutions under CAS.[44] These NERCs thus act as incubators. Accordingly, state allocations to R&D institutions were reduced and autonomy in R&D institutes and mobility of R&D personnel was increased. University cooperation with large and medium size enterprises was also encouraged in order to: (a) update their technologies; and (b) to assimilate the imported technologies. Technology markets were established for effective transfer of technologies. The research institutes were encouraged to commercialize their research results, through these technology markets which acted as inter-mediatory bodies. But the linking of research institutes with the enterprises through technology markets did not meet the expected results.[45]

With the slashed budgets and failure to meet the demands set by enterprises, a number of universities in China have been compelled to market their knowledge and are shifting to manufacturing to finance their research and educational programmes. Many institutes and universities have responded positively to the budget cuts by establishing their own 'spin offs' or 'New Technology Enterprises' (NTEs). This has been considered as one of the major successful outcomes of the S&T reforms. The success of spin-off enterprises and NTEs seems to have legitimized the new commercial role of the research institutes and universities leading to the creation of a new culture of scientific entrepreneurship.

To promote these NTEs and provide a favourable environment for the development of high technologies, High Technology Development Zones (HTDZ)[46] (similar to S&T Parks in developed countries) were promoted through the Torch Programme (TP) initiated by SSTC (now MOST) in August 1988, with the objective to develop new and high-tech products in close cooperation with universities and CAS research institutes. TP helps in commercialization of high tech and new technologies resulting from '863 Programme' and 'National Key Technologies Programme'. The Beijing New Technology Industrial Development Zone, founded in 1988, was the first of its kind in China. So far, it boasts of more than 2000 new and high-tech enterprises employing 50,000 people. In 1992, it realized trade income of 16 billion Yuan and industrial output of 5 billion Yuan. In 1996, it was converted into Zhongguancun Science and Technology Park (ZSTP) having the highest concentration of intellectual resources. It comprises 68 institutions of higher education represented by Bei da and Qinghua universities and 213 scientific research institutions including institutions from CAS. There are also about 36 per cent of the total academicians of CAS and CAE working in this Park. The main focus is on electronics information, optical mechatronics, bioengineering and new medicines, new materials and environment protection. An investigation of 17 colleges and universities in Beijing, Shanghai, Jiangsu and Hubei also showed that during 1991–92, more than 5,000 inventions by college students have been put to use, which are reported to have been received well by the industrial circles, pointing to improvement in academia-industry linkages. By 1992, 52 national-level HTDZs had also been established with 12,900 high/new-tech enterprises. By the end of 1993, some 67 R&D Centres for Engineering Technology were also established with the support of the state to help promotion of transfer of research results to the industry. The main purpose was to further improve R&D for engineering technologies and strengthening the linkages between R&D institutions and enterprises. In addition, 100 enterprises set up their Technology Development Centres (TDC's) during 1993 and 1994.

In 2001, there were 53 such zones with about 20,796 NTEs.[47] About 1,000 of them are affiliated with the educational sector. At present, China has 120 HTDZs at different levels. These zones have been developed with different specialties. For example, the 'Shenyang Zone' features integration of machinery and electronics, while the 'Nanning Zone' focuses on the development of agro-biotechnology. 'Daqing Zone', established in 1992, places high priority on promotion and development of oil and petrochemical equipment and refined chemical technology. This zone enjoys cooperation from Qing Hua University, Beijing, to develop low temperature nuclear heating systems to cope with severe winters. It has also linked itself with Microorganism Research Institute of CAS in respect of a key project which relates to direct pouring of micro-organisms into the strata for raising oil recovery by 30 per cent. The major technologies being developed in these zones relate to new materials, bioengineering, electronics, information technology, integration of machinery and electronics, efficient new energy sources, energy and environmental protection technology. Some high technologies, such as those in the fields of fibre optic communications and electronic information have grown into major industries in Wuhan, Hubei and Beijing HTDZs . In 2000, the total foreign exchange earnings by these 53 HTDZs was US$ 18.6 billion, up 102 times from that in 1991 and 56.1 per cent over 1999 and was expected to rise to US$ 30 billion by the year 2005.[48]

High/New Technologies Pioneer Service Centres (PSCs), also known as 'Incubators', have been established in high-tech zones to transfer scientific research results to production. The first such Centre was founded in Wuhan in Central China in 1987. These centres provide scientific workers a place to turn scientific results into technical applications. By the year 2000, these had hatched 5,293 enterprises. Since 1987, more than 300 incubators have been established in various provinces and cities. Firms and research institutes need only to pay low rents to develop their products and are exempted from tax for two years. Services such as industrial and commercial registration for enterprises, housing lease and facilities for loan are available.[49]

While a number of measures are being taken up to establish linkages between academia and industry and a number of achievements have been cited in the form of 'spin off' enterprises, a lot is still to be done in this direction to create proprietary innovations to build up indigenous technological capability. There is a wide regional disparity too. The eastern region in China, because of its richness in academic personnel and resources, leads in proprietary innovations, the middle and the western regions still need to pay a lot of attention particularly in the high-technology developments.

Transformation of R&D Institutes into Industrial Institutes

In 1998, the State Council decided to abolish 10 ministries (including the Ministries of Machine Building, Metallurgy Industry and the Ministry of Coal, among others) with a view to increasing the power of the market in resources allocation. Meanwhile, the transformation of 242 R&D institutions affiliated to the 10 ministries became one of the key issues; for they employed in 1998, about 115,000 staff, 63,000 of which were scientific persons and another 43,000 were Scientists and Engineers. On 22 February 1999, the Ministry of Science and Technology, the State Commission of Economic and Trade, the State Commission of Development Planning, the Ministry of Finance, and two other government agencies decided that all these 242 government-owned R&D institutions should be transformed completely into S&T enterprises by the end of June 1999 with a view to removing the barriers between research and production and strengthening the national innovation system. These S&T enterprises have been grouped into five categories: Independent technology development research organizations wholly transformed into enterprises; Enterprises with a scope to develop and produce high and new technology products; Academic and scientific research organizations which industrialize independently science and technology results and establish enterprises; High science and technology enterprises in the High and New Technology Development Zones, and Collective science and technology enterprises recognized by the Ministry of Science and Technology.

What defines a science and technology enterprise is its R&D expenditure and a high proportion of technological personnel engaged in high technology products. At the national level, the first batch of 242 research-institutes affiliated to the former State Committee for Economy and Trade and the second batch of research-institutes affiliated to other ministries had been transformed into enterprises by the end of 2001 (Table 2.8).

The restructuring has been implemented in three phases: (*a*) Restructuring 242 industry-oriented research institutions; (*b*) Restructuring 134 research institutions affiliated to different governmental industrial

Table 2.8

Transformation of Public R&D Institutions in China After 1999

Transformation Year	Number of Transformed R&D Institutions	Owners of the Transformed R&D Institutions	Status After Transformation	Preliminary Result of the Survey by the MOST in May 2002 on 290 Transformed R&D Institutions
1999	242	Ex-State Economy and Trade Commission	Enterprises	• Revenue in 2001: 1.5 times of in 1999; Profit in 2001: 2.6 times of in 1999; Tax in 2001: 1.9 times of in 1999. R&D expenditure annual increase rate in 2001: 16.2%; in 2000: 6.84%.
2000	134	11 Ministries: Ministry of Construction, etc.		• Patent application annual increase rate in 2001: 9.6%. Employee average salary in 2001: 142.6% of that in 1999.
1999–2002	660	Local Governments		• 92.6% of them set up enterprises accounting system; 88.65% entered the local unemployment insurance; over 10 of them went public in the stock market
2001	98	4 Ministries and Agencies: Ministry of Land and Resources, etc.	89 institutions: Non-profit Organizations 61 institutions:	
2002	107	9 Ministries and Agencies: Ministry of Agriculture, etc.	Enterprises Others: Merged into universities,	
2004	43	5 Ministries and Agencies: Ministry of Health etc.	transformed into intermediary organizations	

Source: Li, Xueyong et al. (2002), The Address of Vice President of MOST, Li Xueyong, in the Conference on Public R&D Institutions Reform (in Chinese), 21 October 2002, Beijiing, P. R. China. Available at http://www.most.gov.cn/Attach/att_ZCYXGG_17_2.doc (Accessed on 1 April 2004).

departments; (c) Restructuring 265 research institutions of public goods. These 242 of the larger scientific research institutes under the control of the government have now been successfully transformed, mostly into private companies and a few into technical services institutions. Among these, 10 are now traded publicly on the stock market. In 2002, these 242 transformed research institutes received 676 million Yuan and took over the projects of national S&T programmes. These earned a profit of 1.1 billion Yuan from the sales revenue of 9.1 billion Yuan. According to a survey conducted by Ministry of Science and Technology (MOST) in 2002 on 290 transformed enterprises showed that these transformed research institutes play very important roles in promoting progress of industrial technology as shown in Table 2.8.

Following the transformation of 242 state-owned independent research institutes, the CAS has also transformed its institutes exclusively engaged in technological development into self-supporting enterprises. By the end of 2001, 13 CAS affiliated institutions had already completed this institutional transition, including the Chengdu Institute of Computer Application, the Chengdu Institute of Organic Chemistry, the Guangzhou Institute of Electronic Technology, the Guangzhou Institute of Chemistry, the Shenyang Institute of Computing technology, the Beijing Research Center of Software and five Research Centres for Scientific Instruments in Beijing, Shenyang, Chengdu, Xinxiang and Nanjing.[50] These converted R&D institutes changed their organizational structure from being subordinate to the government to become competitive players in the market as independent legal entities.

During the tenth five-year period (2001–05), China made further progress in restructuring its S&T management system. Aiming at turning R&D institutes into S&T-oriented enterprises, the efforts have resulted in institutes enjoying an increasingly enhanced innovation capability. In 2004, the patent applications filed by these transformed institutes saw a growth of 22.06 per cent, with 1,150 patents granted, or 42.68 per cent up compared with the preceding year, of which invention patent grants enjoyed a growth of 57.74 per cent. Statistics show that in 2004, these institutes produced a combined revenue of RMB 45 billion, 95 per cent up as compared with 2000.[51]

However, the transformation of these R&D institutions also has some negative impact on technological innovation capability in China[52] for the following reasons:

1) The model of the transformation is too simple, and neglects the diversity of institutes in nature and the variety of social economic demands. In practice, most of the 242 state-owned independent research institutions have been transformed into state-owned enterprises, which still need to be transformed into modern corporations or companies with mixed ownership so as to overcome the problems such as low efficiency that state-owned enterprises usually have. The technological innovation capacity of the enterprises in general is still weak.

2) Some of these institutions provide public goods for industrial sectors, not for specific enterprises. They cannot obtain enough economic returns from the market to support themselves. Most Chinese enterprises are weak in technological innovation, and small in business scale (compared with world's leading MNCs). Furthermore they cannot afford to invest large amounts of money in R&D even if they realize that R&D is extremely important for their competitiveness.

3) Some preferential policies to promote the transformation have to be adjusted or even abolished after China's entry into the WTO. For example, these transformed institutes are no longer R&D institutes, but are still treated as state-owned R&D institutions. They cannot obtain enough economic returns from their research in the market and need government support.

The Funding System for S&T

Over the years, there has been considerable increase in the total expenditure on R&D, which increased from 12.543 billion Yuan (US$ 2.65 billion) in 1990 to 89.60 billion Yuan (US$ 10.70 billion) in 2000 which was 1.01 per cent of GNP, but it was still well below the target of 1.5 per cent set in 1995 for the ninth Five-Year Plan (1996–2000). In 2010, it further increased to 706.26 billion Yuan (US $104.48 billion) which was 1.76 per cent of GNP. During the 11th Five-Year Plan (2006–10), the target set was 2 per cent (Table 2.9).

Table 2.9

Total Expenditure on R&D in China (1990–2010)

	Yuan (In Billions)	US$ (In Billions)	% of GNP
1990	12.543	2.65	0.71
1995	28.6	3.45	0.50
2000	89.60	10.70	1.01
2001	104.25	12.59	1.09
2002	128.76	15.53	1.1
2003	153.96	18.57	1.31
2004	196.63	23.75	1.23
2005	245.00	29.84	1.34
2006	300.30	37.54	1.42
2007	368.5	49.13	1.49
2008	461.6	65.94	1.54
2009	580.21	85.08	1.70
2010	706.26	104.48	1.76

Sources: Various issues of Beijing Review, China Daily and China Science and Technology Newsletter. China Statistical Yearbook, 2009, National Bureau of Statistics of China, Beijing.

The Interim Regulations of the State Council for the Administration of S&T appropriations issued on 23 June 1986 provided that 'research projects of national priority' were to remain under the control of the State Plan, while other activities concerning 'technology development and applied research projects' were to be managed by economic levers and market regulations. With this policy, over the years, the operating expenses provided by the State for 'technology development and applied research projects' are being gradually reduced with the aim of making these institutes basically self-reliant in their operating budgets. R&D institutions can sign R&D contracts directly with the enterprises, set up research–production entities or provide consultancy services. In 1990, the state contributed 54.9 per cent of the total R&D funds, industry 23.4 per cent and financial institutions and others 21.7 per cent. By the turn of the century, the contribution of enterprises to

R&D increased to 60 per cent and in 2010 it further increased to 73.4 per cent (Table 2.10).

Table 2.10

Sector-wise Contribution to R&D Expenditure in China (%), 1995–2010

Sector/year	1995	2000	2004	2005	2006	2007	2008	2010
Industrial Sector	43.7	60.0	66.8	68.2	71.1	72.8	73.3	73.4
Government Research Institutions	42.1	28.8	24.8	20.9	18.9	18.7	17.6	16.8
University Sector	14.3	11.2	8.4	9.9	10	8.5	8.5	8.5

Source: Various issues of China S&T Newsletter, China S&T Statistics, Data book, Ministry of Science and Technology, PRC.

Another important source of funding Beijing utilized is 'venture capital' to support the development of high technologies and those involving greater risks. Banks are also being actively involved to provide loans for S&T work. In fact, S&T loans from banks increased from 15 million Yuan in 1984 to 6.3 billion Yuan in 1993.

According to the nature of the research work, in 1990, 7.3 per cent of R&D funds were spent on basic research, 28.5 per cent on applied research and 64.2 per cent on experimental development. In 2008, this proportion was 4.8, 12.5 and 82.7 per cent respectively, showing enormous increase in innovation activities to speed up technological developments. Ironically, the share of R&D expenditure on basic research has been reduced from 12.4 per cent in 1986 to 7.3 per cent in 1990 to 4.8 per cent in 2008 (Table 2.11) obviously in favour of emphasis on short term technological innovation achievements.

In 2010, special equipment manufacturing sector claimed the highest R&D expenditure intensity (as a proportion of main business income) by 2.04 per cent. The sectors enjoying an R&D expenditure intensity between 1.5 per cent and 2 per cent are pharmaceuticals (1.82 per cent), generic equipment (1.59 per cent), electrical machinery and equipment (1.59 per cent), and instruments, office machinery (1.50 per cent).[53]

From the foregoing it is clear that while there has been a 20 times increase in R&D since 1994, it is still considered to be marginal when compared to international levels of 2.5–3 per cent of GNP by the advanced countries. In 1994, in the USA and Japan, total expenditure on R&D was

Table 2.11

Discipline-wise Contribution of R&D Expenditure in China (%), 1995–2008

Discipline/year	1990	1995	2000	2005	2006	2007	2008
Basic Research	7.3	5.2	5.7	5.4	5.2	4.7	4.8
Applied Research	28.5	26.4	19.2	17.7	16.8	13.3	12.5
Experimental Development	64.2	68.4	75.1	76.9	78.0	82	82.8

Sources: Various issues of China S&T Newsletter, China S&T Statistics, Data book, 2003–07, Ministry of Science and Technology, PRC, China Statistical Yearbook 2009, National Bureau of Statistics of China, Beijing.

US$ 200 billion and US$ 120 billion respectively compared with China's US $65.94 billion in 2008. With such a low national R&D expenditure, international competition is out of the question in the face of most of the MNCs from advanced countries like General Motors, Siemens, Ford and Hitachi, each of which spends between US$ 4 to 6 billion annually on R&D. R&D institutes must have some critical human and financial resources to generate tangible research results, which seem to be lacking in the Chinese system, although enormous efforts are being made in this direction by concentrating the available resources.

The current level of expenditure on basic research is around 4.8 per cent of the total state S&T allocations, with the CAS and the higher educational sector accounting for 90 per cent of the national basic research funds. This has also resulted in wide regional disparity between the central and local institutions. In 1990, national level institutes numbered 1,123, that is, 22.8 per cent of similar organizations in the country. Funding for such institutes was 5.764 billion Yuan, i.e., 82.5 per cent of the total funding for S&T development in the country. S&T personnel in the national level institutes numbered 0.397 million, that is, 68.2 per cent of the total in the country. These figures reveal that the number of institutes at the national level is small, but they have a major share in terms of investment resources. These research institutes at the national level are the main force in S&T development in the country. The local sector, thus, is poorly funded and equipped. Some of the research institutes are so small that they should not be listed as R&D units per se. For example, one-fourth of them have less than 20 S&T personnel of which only one or two may be scientists or engineers.

Moving towards Innovation-oriented Country

As at present, China is in the process of building up National Innovation System (NIS)[54] to rejuvenate Chinese economy through science, education and sustainable development. The knowledge-based industry has become a new focus of economic growth. While the ninth Five-Year Plan (1995–2000) mainly focused on the creation and expansion of basic and high technology research facilities through restructuring and increasing resources in the CAS institutions and the universities,[55] the tenth Five-Year Plan (2000–05) basically focused on strengthening linkages among various institutions under the CAS and institutions of higher education and the enterprises by various mechanisms like restructuring its S&T management system to create S&T Enterprises and putting more stress on innovative activities, at the same time reducing the financial burden on the State for innovations and technological development. The eleventh Five-Year Plan (2006–10) proposed one step further to concentrate on networking of institutions and create a National System of Innovation to make proprietary innovation and make China an innovation-oriented country.

The Chinese Ministry of Science and Technology, in collaboration with other government agencies, including National Development and Reform Commission, Ministry of Finance, Ministry of Education, Chinese Academy of Sciences, Chinese Academy of Engineering, National Natural Science Foundation, China Association for Science and Technology, and State Administration of Science, Technology and Industry for National Defense, released a national science and technology development plan for the twelfth five-year period (2011–15). This plan defines the following main objectives: enhancing China's proprietary innovation capacity, competitiveness, and international influence in the area of science and technology; striving for major technological breakthroughs in the key areas; providing strong support to the change of economic development mode; and establishing a national innovation system featured with clearly-defined functions, rational structures, sound interactions and efficient operations. China will work hard to sit in the 18th place rather than in the current 21st place in the world, in terms of comprehensive innovation capacity.

Seven Chinese government agencies, including the Ministry of Science and Technology, recently released a national medium- and long-term plan (2010–20) for the development of human resources in the area of science and technology. It says, in the coming decade China will strive to train and bring out 3,000 young talented leaders for the development of cutting edge technologies and strategic emerging industries. The plan says China will make young scientists' proprietary research activities part of the national science and technology projects, through the mode of 'talents + projects', in an effort to enhance the capacity building of researcher contingents and young academic leaders, under the combined mode of strengthening the existing team of academic leaders, while reserving and developing the new one. Efforts shall be made to support young scientists to be independent leaders of research projects, reforming the current S&T programme management system and securing more support to young scientists. Meanwhile, tilted support will be given to the research activities under the independent leadership of outstanding young scientists under the age of 35. China will expand its S&T personnel contingent. People who are engaged in R&D activities will be raised from 1.965 million person-year in 2008 to 3.80 million person-year in 2020, and R&D personnel will be raised from 1.05 million person-year in 2008 to 2 million person-year in 2010. By 2020, China's per capita R&D expenditure enjoyed by R&D personnel will hit Yuan 1 million/year from Yuan 277,000/year in 2010, reaching the level of a moderately developed country. In the coming decade, China will work hard to build six major contingents of scientists having the strength of original innovation, researchers with strong innovation capability, engineers, young academic leaders, innovative entrepreneurs, and S&T management.[56]

To boost the innovation system, the patent system was further amended[57] in 2008 to strengthen proprietary innovations and attract foreign multinational companies to invest in R&D in China. Team research with multidisciplinary approach with institutional networking between academia, enterprises, financial institutions to promote venture capital, and the government became important to exploit the specialized knowledge to seek specific and strategic applications.

The building of NIS thus throws a new challenge to the Chinese scientific community in terms of attaining desired results. So far the ongoing reforms seem to have a positive impact on revitalizing the

Table 2.12

Invention Patents Granted in China, 1985–2008

Year	Total	Domestic	Percentage	Foreign	Percentage
1985–92	15773	6072	38.5	9701	61.5
1993	6556	2634	40.2	3922	59.8
1994	3883	1659	42.7	2224	57.3
1995	3393	1530	45.0	1863	55.0
1996	2977	1395	46.9	1582	53.1
1997	3494	1532	43.8	1962	56.2
1998	4733	1655	35.0	3078	65.0
1999	7637	3097	40.6	4540	59.4
2000	12683	6177	48.7	6506	51.3
2001	16297	5388	33.0	10909	67.0
2002	21476	5854	27.3	15622	72.7
2003	37154	11404	30.7	25750	69.3
2005	53305	20705	38.8	32600	61.2
2006	57786	25077	43.4	32709	56.6
2007	67948	31945	47.0	36003	53.0
2008	93706	46590	49.7	47116	50.3

Sources: MOST, China Science and Technology Indicators, 2000; 2001; State Intellectual Property Office (SIPO), Beijing, Statistics for 2001–03 (in Chinese); Wang Jun (2004), China Statistical Yearbook, 2009. Beijing, National Bureau of Statistics of China. China S&T Statistics—Data Book, 2007, Ministry of Science and Technology of PRC. China Statistical Yearbook, 2009, National Bureau of Statistics of China, Beijing.

Chinese scientific community as is indicated by the output of papers for publication and patenting. There has been remarkable growth in patent activity as shown in Table 2.12. The granting of domestic patents in the year 2008 have increased more than seven times since the year 2000; so also is the case in the granting of foreign patents. Similarly, there has been a remarkable increase in scientific publications as indicated in Table 2.5. The world rank of Chinese publications shifted from the 10th position in 1995 to the 8th position in 1999, to the 5th in 2003 after the US, Japan, UK and Germany. Zhong Xiwei and Yang Xiang Dong, in their study on the analysis of S&T reforms on the National Innovation System, conclude that S&T reforms in general have been effective in motivating universities and research institutions (URIs) in building up innovative capacities of enterprises by promoting University-Research Institutions-Industry linkages. But still a lot is to be done to strengthen these linkages.[58] A study carried by OECD also points to the Chinese innovation system still not being fully developed and still imperfectly integrated with many linkages between actors and sub-actors (regional versus national) remaining weak.[59]

Building indigenous technological innovation capability at par with Japan and Korea appears to be the most important task facing China. According to the Global innovation index, China's rank is still 43, far behind the US (11) and UK (14). For the Chinese, this is a major problem in the wake of acute scarcity of expert S&T manpower, funding, regional disparity, large number of enterprises still lacking linkages and political interference in academic institutions. Quality of research publications is also wanting as shown by a report from Science Watch, according to which the percentage of total papers among the top 1 per cent most cited papers in all fields, China stands at 0.5 per cent as compared to the US 1.87 per cent, UK 1.53 per cent and Germany at 1.27 per cent. Gross expenditure on R&D as a percentage of GNP increased marginally from 0.65 in 1987 to 0.70 in 1992 but it declined to 0.5 per cent in 1994. In the 9th Plan (1995–2000), the objective was to raise the total expenditure on R&D to 1.5 per cent of GDP and raise the expenditure on basic research from 7 per cent to 10 per cent, but by the end of the plan in the year 2000, the total R&D expenditure was still only 1.01 percent of GDP and expenditure on basic research, in fact, decreased to 4.8 per cent.

On the whole, in spite of intensive reforms during the past decade, there has not been any remarkable change in the technological backwardness of most of the industries, technical processes and managerial skills. Statistics show that more than 93 per cent of the nearly US$ 100 billion of foreign investments have ended up in general processing industries. No substantial capital has been invested in high technology industries. Moreover, most of the research institutions continue to work in isolation instead of working in collaboration with enterprises. More than 30,000 laboratory research results are registered every year but only 10 per cent of them go into production. Extensive in-house R&D efforts by industry are wanting. According to Jiang Lu'an, Director of the Technology and Equipment Department of the State Economic and Trade Commission, "Only half of the country's large and medium enterprises, which are described as pillars of China's economy, have their own R&D institutes compared to 100% in USA and Japan". The core of the reforms still reiterates "to establish a new operational mechanism, combine improved planning and management with strengthened market regulations in an organic way".

Conclusions

The developments narrated above clearly indicate the changing structural and institutional context of scientific research in China in an entrepreneurial mode with increased networking, partnerships and collaborations with a multidisciplinary approach to generate specialized knowledge for specific applications leading to proprietary knowledge generation and innovations. The new environment of global competition requires a situation where the norms of generation of new knowledge are as important as the norms for innovation and commercialization, and it has put the scientific community in a dual role. One must note, in particular, that China has been able to achieve this in a largely state-controlled environment, where political will and commitments have played an increasingly important role. The main factors responsible for Chinese success are quick implementation of various initiatives by the government, supervision by the Party committees at the various levels of institutions and heavy decentralized approach with commitment and strong supervision. Other relevant factors are a more open environment for import of advanced technologies, better tax incentives and promotion of MNCs, FDI and investment in R&D, greater impetus to the SMEs and the creation of better linkages between various institutions and technology absorption and adaptation.

While India's success in the scientific and technological domains cannot be overlooked, it is still caught in a paradox of too much democracy at all levels of social and economic activity. It needs a big political push and inclusive decentralized approach to realize its well defined plans and objectives to make its international presence felt.

Bibliography

Chen Zhi (2004), Analysis of the Main Characteristics of Transformed Science and Technology Enterprise (*zhuanzhi kejixing qiye de zhuyao thezheng fenxi* [in Chinese]), *Zhongguo keji luntan* (China Science and Technology Forum), January, pp. 71–74. http://www.chinaeducenter.com/en/cedu/ceduproject211.php (accessed on 14 October 2010).

The Economic Times (2010), CII-INSEAD Innovation Index 2009–10, *The Economic Times*, 24 June.

Cui Ning (1999), 'Top Science Honour with US $ 602,400', *China Daily HKE*, 10 August.

Guan J. C., Richard C.M. Yam, E.P.Y Tang and A. K. W. Lau (2009), 'Innovation Strategy and Performance during Economic Transition: Evidence in Beijing, China', *Research Policy* 38, pp. 802–12.

Hu Jintao (2009), 'Congratulations on the 60th Anniversary of CAS', *CAS Newsletter*, October, http://www.conference.ac.cn/Newewsletter/newsletter68-0.html (accessed on 13 October 2010).

Kharbanda, V. P. (2012), 'India Should Heed but Not Fear China's Science', SciDev.Net, 23 February.

Ren Shengli and Ronald Rousseau (2002), 'International Visibility of Chinese Scientific Journals', *Scientometrics* 53, no. 3, pp. 389–405.

'Scientific and Technological Sector', http://202.84.17.11/english/china-abc/science.htm (accessed on 28 March 2002).

Teng Hsiao-Ping (1978), 'Speech at the Opening Ceremony of National Science Conference', *Peking Review* 21 no. 12, pp. 9–18.

'Which Countries Publish the Most' (2007), *Science Watch* (May/June).

Wu, Yuanli and Robert B. Sheeks (1970), *The Organization and Support of Scientific Research and Development in Mainland China*. New York: Praeger Publishers.

Zhou Qiong and Yang Aili (2010), 'Fabricated High-Tech Boom: Tax Incentives for High-Tech Forms Create Huge Application-Processing Industry', Caixin Online, www.marketwatch.com/story/chinas-high-tech-subsidies-boosts-paper-pushers-10.08.2010 (accessed on 29 September 2010).

Notes and References

1. Zhao Ziyang, Speech to the National Science and Technological Work Conference, 6 March 1985. Xinhua in Chinese, 20 March 1985. SWB/I:-' E/7908/3II/4.
2. Ibid.
3. Cong Cao, 'The Chinese Academy of Sciences: The Selection of Scientists into Elite Group', *Minerva* 36, no. 4 (1998): 323–46.

4. Cong Cao and Richard P. Suttmeier (1999), 'China's "BRAIN BANK": Leadership and Elitism in Chinese Science and Engineering', *Asian Survey*, 39, no. 3 (1999): 525–59.

5. Yen-Farn Wang (1991). 'China's Science and Technology Policy: 1949–1989', *Stockholm Studies in Politics*, 39 (1999), Stockholm: University of Stockholm.

6. The number of publications authored by Chinese scientists in international journals has increased over the years, ranking 20th in 1985 in the world raised to 11th in 1996 and 9th in 1997, to 5th place in 2003 after US, Japan, UK and Germany. While China's scientific papers rose 15 times from 6,509 in 1990 to 94,800 in 2007, India could only increase output from 10,103 papers in 1990 to 30,000 in 2007. India also lags far behind China in the number of patents granted: in China patents increased from 12,683 in 2000 to 93,706 in 2008, while in India these rose from 1,318 in 2000 to only 7,539 in 2006 (see V.P. Kharbanda, 'India Should Heed but Not Fear China's Science', *SciDev.Net*, 23 February 2012).

7. Kathleen G. Dugan, Su Dajun and Yang Ji (eds), Science and Technology in China: Selections from the Bulletin of the Chinese Academy of Science (Beijing: International Academic Publisher, 1988).

8. CAS Journal (1989), 'Overview of a Decade's Reform in CAS', 14(3).

9. 'Revised By-laws on National Lab Management', *China Science and Technology Newsletter* (CSTN, MOST), (10 September 2008): 522.

10. Legend was the first company to introduce the PC concept in the People's Republic of China in 1984. Since 1997 the company has been the leading PC brand in and around China with annual revenues (as of May 2005) of approximately US$3 billion. In 2003, the Legend Group Limited launched its new brand Lenovo to cater to the group's future business development and laid the groundwork for its expansion into overseas market. (See 'Lenovo Bridging East and West to Build a Global Brand', http://www.pfoertsch.com/wiki/index.php/Main/LenovoCase (Accessed on 7 October 2010).

11. 'More Key Research Facilities', *China Science and Technology Newsletter* (CSTN, MOST) (30 March 2008): 506.

12. Cong Cao and Richard P. Suttmeier, 'China's "BRAIN BANK"', 534.

13. PPKIP was formally initiated in 1998 and was completed by the year 2010. It was divided into three phases: The initial phase from 1998 to 2000; the implementing phase from 2000 to 2005 and Optimizing phase from 2006 to 2010. The main objective is to restructure about 80 national institutes with powerful innovation capabilities; 30 of which will become internationally acknowledged high level research institutions; and 3 of them will be first class in the world. A vigorous training system for the advanced S&T talents will be formed. To build the CAS into a prominent base of modern science and innovation culture and make it a major representative of China in the international science community through further opening to the outside world. The initial phase was completed in 2000.

14. *China Science and Technology Newsletter* (CSTN, MOST), (10 October 2005): 417.
15. Huang Shiqi, *University Research in China* (1988), 21.
16. M. Roche, 'Some Facts and Many Impressions on Science in People's Republic of China', *Interciencia* 5, no. 2 (1980): 113 (in Spanish).
17. Zelda F. Gamson, 'After the Revolution comes the Educational Testing Service: Notes on Higher Education in China' (University of Michigan, 1979), 10–11.
18. 'China's Higher Education for the Next Century' (Beijing; Xincheng chubanshe chu, 1995), 19.
19. China Education Centre, 2004, 'Project 211' and 'Project 985'.
20. V. P. Kharbanda, 'Scientific Communities in India and China: Formation, Growth and Changing Structure' (Saarbrucken, Germany: VDM Verlag Dr. Müller AG & Co. KG, 2009), (www.vdm-verlag.de).
21. 'China's Higher Education for the Next Century'.
22. Huang Wei, 'Reforms of the College Remuneration System', *Beijing Review* 43, no. 3 (2000): 16–19.
23. Cui Lili, 'The Metamorphosis of Beijing University', *Beijing Review* 36, no. 13 (1993): 19–22.
24. 'China's Higher Education for the Next Century'.
25. *China Science and Technology Newsletter* (CSTN, MOST), 63 (15 December 1995): 3.
26. *China Daily* (30 October 1996), 4.
27. 'New University Science Park Plan', *China S&T Newsletter* (10 September 2011): 630.
28. Li Li, 'A Tall Order'. *Beijing Review* 53 no. 22 (2010): 18–19.
29. Yao Bin, 'Seeking Educational Excellence', *Beijing Review* 53, no. 22 (2010): 2.
30. V. P. Kharbanda, 'Scientific Communities in India and China'.
31. On the basis of a report prepared by the Chinese Ministry of Education on the enrolment of candidates in the universities.
32. 'Twelve Projects Staged to attract Talents', *China S&T Newsletter* (20 November 2011): 637.
33. Hua Kuofeng, 'Unite and Strive to Build a Modern Powerful Socialist Country: Report on the Work of the Government (First Session of the 5th National People's Congress, 26 February 1978), *Peking Review* 21, no. 10 (1978): 7–40.
34. Jia He Peng, 'Outstanding Scientists to Receive Rewards', *China Daily* (22 October 1999), 2.
35. See 'Science and Technology System', www.edu.cn (Accessed on 7 December 2001).
36. According to Baark, there were 5,000 S&T journals in China during the early 1980s. See Erik Baark, 'China's Information Services: A Brief Appraisal of Structure and Policy Issues', *NISSAT Newsletter* 4, no. 1 (January–March 1982): 4-8; 4 no. 2 (April–June 1982): 18–24.

37. Ren Shengli, Ping Liang and Zu Guangan, 'The Challenge for Chinese Scientific Journals', *Science* 286 (26 November 1999): 1683.

38. V. P. Kharbanda, 'Scientific Communities in India and China'.

39. Basic Statistics on Science and Technological Activities of China Association for S&T (1998). http://www.edu.cn (accessed on 7 February 2001); *Science* 286 (26 November 1999):1683, quoted in Science Policy Information News (SPIN Website) (accessed on 29 January 2000).

40. In 2002, the number of research paper authored by Chinese scientists in SCI further increased to 40,758, advancing its position from 20th in 1985 to the 6th place in the world, after US, Japan, UK, Germany and France.

41. 'China's 5th Place for World S&T Papers', *China Science and Technology Newsletter* (CSTN, MOST),(2003): 2.

42. 'Government Pushes to Encourage Scientific Innovations', New China News Agency (NCNA) (24 August 1999), SWB/FE/3621G/5.

43. Fang Xin, 'Progress of Science and Technology Reforms in China' in *Science and Technology Strategies for Development in India and China: A Comparative Study*, eds V. P. Kharbanda and Ashok Jain (New Delhi; Har Anand Publishers, 1999), 159.

44. Mu Rongping and Lian Yanhua, 'Reform on Science and Technology System in China: Case study on National Engineering Research Centers', paper presented in the 1997 ICSPS Symposium on Science, Technology and Innovation Policy at the Turn of the Century: Experiences and Prospects, organized by the Institute of Policy and Management, Beijing, China.

45. Gu Shulin,*China's Industrial Technology: Market Reform and Organisational Change* (London; Routledge, 1999), 25.

46. These HTDZs are essentially Science and Technology Parks (STPs) and are an industrial complex, close to the places of learning like universities, colleges or polytechnics or research laboratory having formal or informal links. These are designed to encourage formation of knowledge-based industries in a high quality and competitive environment. An STP has a management function for transfer of technology and business skills to enterprises on site; aims to reduce the time gap between scientific invention and its commercial application. The first STP was established in Stanford University, USA around 1950. However, the STP movement picked up only in the late 1980s. At present there are 200 parks in the USA, 52 in the UK, 86 in Germany, 27 technopolis in Japan, and 120 high-tech zones in China. In UK, the 52 parks have come together to form a Science Park Association (See B. M. Naik and W. S. Kandhikar, *Higher and Technical Education: Book of Knowledge* [New Delhi: Gyan Publishing House, 2010], 203). Beijing Technological Development Pilot Project was approved in 1988. In China the first batch of its kind of 26 HTDZs were approved by the State Council in 1991. In 1992, another batch of 25 was approved. In 1997, the Yangling Agricultural High Tech Region was approved. In 2001, there were a total of 53 HTDZs with 20,796 NTEs (See Li Mu, 'High-Tech Development Zone Flourish', *Beijing Review* 44, no. 24 (2001): 21–22).

47. According to Suttmeier and Cao, there were 65,000 NTEs in these 53 HTDZs in various parts of the country (Richard P. Suttmeier and Cong Cao (1999), 'China Faces the New Industrial Revolution: Achievement and Uncertainty in the Search for Research and Innovation Strategies', *Asian Perspective* 23 no. 3 [1999]: 153–200, 162). See also Li Mu, 'High-Tech Development Zone Flourish', 21.
48. Li Mu, 'High-Tech Development Zone Flourish', 21.
49. However, Tang reported only 150 incubators in April 2001 (Tang Yuankai, 'Full Speed Ahead on High-Tech Development', *Beijing Review* 44, no. 17 (2001): 22–24, 23). See also Wei Liming, 'Torch Program Making Rapid Progress', *Beijing Review* 36, no. 10 (1993): 18–23.
50. Mu Rong Ping, 'The Impact of R&D Institute Reform on Technological Innovation in China', seminar presentations (Creation of Innovation: Research Centers, Institutes and Universities, 2003), http://www.law.gmu.edu/nctl/stpp/us_china_pubs/technical_innovation_bc_english.pdf (accessed on 16 October 2010).
51. 'Transformed Institutes With Better Innovations', *China Science And Technology Newsletter* (CSTN, MOST),(28 February 2006): 431.
52. Mu Rong Ping, 'The Impact of R&D Institute Reform on Technological Innovation in China'.
53. 'China's 2010 S&T expenditure released', *China S&T Newsletter* (30 September 2011): 632.
54. The China National Innovation System initiated in 1998 is a networking system composed of institutions involved in knowledge innovation and technology innovation. It includes the following: knowledge innovation system netted with the state research institutions and key teaching research universities; technology innovation and technology application system with industrial enterprises; knowledge dissemination system with schools and universities conferring different levels of degrees and with lifelong education open to the public. In this connection, the national institutes play a unique role in the popularization of science among the people in science and technology development and expansion, and in the promotion of scientific methodology.
55. The 1995 conference announced the decision 'to accelerate scientific and technological progress' through the strategy of 'revitalizing China through science and education'. It also called for 'stabilizing one end and freeing up all the rest' which meant to maintain a small but highly trained research task force to work on basic, applied and high technology research, on the one hand, and to give free play to market mechanisms, on the other. Song Jian, State Counsellor and Chairman of the SSTC, had called for an increase in R&D expenditure to 1.5 per cent of GNP and the share of basic research from 7.0 per cent to 10 per cent by the end of 2000.
56. '3000 Young Leaders for Next Decade', *China S&T* newsletter (20 August 2011): 628.

57. The 'Chinese Patent Law' came into force from 1 April 1985. The first amendment in 1992 added pharmaceutical compositions to the list of patentable subject matters and inaugurated China's membership in the Patent Cooperation Treaty (PCT). The second amendment in 2000 brought China's Patent Law, in line with WTO, into compliance with the Trade-Related Aspects of Intellectual Property Rights (TRIPS) agreement. The third 2008 revision, which came into force on 1 October 2009 is expected to bring the law and processes up to international standards and best practices. With this, Chinese Patent law ensures better protection for the intellectual property rights, has been harmonized with international patent treaties, and involves the 'absolute novelty' standard that is applied internationally.

58. Zhong Xiwei and Yang Xiangdong, 'Science and Technology Policy Reforms and Its Impact on China's National Innovation System', *Technology and Society* 29, no. 3 (2007): 317–25.

59. OECD, *OECD Reviews on Innovation Policy* (China. Paris: OECD, 2008), 651.

3

China in Space*

U.R. Rao

China's military posture, its mastery and level of space technology development, ambition to establish itself as the most powerful space power and its economic advancement which enables it to spend huge amounts of money on development of space technology has serious implication for our country.

A greater synergy must be established between our defence establishments and civilian establishments of space and nuclear energy and ensure the full intellectual strength of our country is employed in planning and execution of our national strategy.

Introduction

Historically China experimented with 'fire arrows', well before 1,000 AD, which in 1045 were converted into lethal solid fuel rockets using sulphur, saltpetre and charcoal. These were effectively used to repel Mongol invaders in the battle of Kaifung-fu in 1232. Old records indicate that these rockets carried iron shrapnel and had used iron-pot combustion chambers for the first time.

---

* The information presented in this chapter is collected from a very large number of published sources including newspapers, government publications, review articles, press releases and personal discussions. The interpretations and conclusions, however, are drawn by the author.

Rocket technology in India started with Hyder Ali who introduced metal cylinders to hold rocket powder fuel and made rockets weighing 5.5 kilograms with a range of 2.5 kilometres. Hyder Ali built a rocket force of 5,000 men, who succeeded in defeating General Coote of British East India Company army in 1780.

Tipu Sultan, who became the Ruler of Mysore in 1782, further improved these rockets by attaching long bamboos to the rockets. These were successfully used against the British in the two Anglo-Mysore wars at Srirangapatnam in 1792 and 1799. Several of these rockets were seized by the British after the defeat of Tipu in the third war, two of which were taken by Dr Congreve and are still preserved in the British War Museum in London. Even though on my personal request the British Museum lent us one of the rockets for over a week, which was exhibited at Bangalore during the IAF Congress in 1988, very few people came to see the historical rocket in spite of the wide publicity given.

China's space programme, which is directed by the Chinese National Space Administration (CNSA), started in the 1950s. China established its first Missile and Rocket Research Centre in 1956. It began its rocket launching in 1960 with sounding rockets. China's early space programme was greatly helped by Qian Xuesen, a cold-war spy deported from the US in the 1950s. Qian Xuesen (1911–2009), who received his Doctorate from the California Institute of Technology, was a genius and is widely regarded as the father of Chinese Missile Programme. Substantial help received from the Soviet Union till 1960, resulted in the design and development of its Long March rocket, which carried China's first satellite Dong Fang Hong-1 weighing 170 kilograms in 1970, making China the fifth country to send a satellite into orbit. Contribution of Qian to the space and missile development, and the support from the Soviet Union are widely credited for the successful atomic test in 1964. Much of the technology behind Shenzhou rockets, which started operating since 1990, are credited to Qian Xuesen.

It is interesting to compare the early history of space technology development in China and India. China established its rocket research centre in 1956. India started its rocket centre in 1963. China launched its first satellite weighing 170 kilograms in 1970 with their own launcher. We launched our first satellite, Aryabhata, weighing 420 kilograms in 1975 but using Russian rocket. It was only in 1980 that we launched a 40 kilograms Rohini Satellite with our own SLV-3 launcher. In terms of

Table 3.1

Historical Developments in Space

Sl. No.		China	India
1.	Establishment of Rocket Research Centre	1956	1963
2.	Launch of Sounding Rockets	1960	1964
3.	Launch of first Satellite	1970 (170 kg)	1975 (420 kg)
4.	Launch of first Satellite with indigeneous launcher	1970 (170 kg)	1980 (40 kg)
5.	Launch of first operational Communication Satellite	1990 (own launcher)	1983 (Shuttle)
6.	Launch of first operational Earth Observation Satellite (CBERS- IRS)	1999 (own launcher)	1988 (VOSTOK)

satellite technology, we launched our highly sophisticated commercial satellite INSAT in 1983 as against China in 1990. Likewise our operational remote sensing satellite was launched in 1988 from Baikonur using a Soviet Rocket as against China's first remote sensing satellite CBERS-1 in 1999. Table 3.1 summarises the historical development of space technology in China and India.

Even though the Cultural Revolution under Mao considerably slowed down China's space programme, it was rectified in the 1980s after Mao's death. It got a big boost particularly after Jiang Zamin, who was a renowned technical expert in electronics, took over as President of China in the 1990s.

Launch Infrastructure

China's Spaceports

China has four major launch sites (see Figure 3.1) for launching scientific, application and military satellites as well as for carrying out missile operations:

1. Jiuquan launch site in the Northern Gobi Desert in far-western China, not far from Mongolia, which is the major launch site for manned flights.*

---

* Figures 3.1, 3.2 and 3.3 are at the end of this chapter.

2. Taiyuan launch site in Shanxi Province, 60 miles southwest of Beijing, in north-central China.
3. Xichang launch site in an isolated corner of Sichuan province in Southern China, not far from Myanmar, Laos and Vietnam.
4. Hainan Island off the southern coast of China, separating the South China Sea from the Gulf of Tongking. Hainan being just 19 degrees north of the equator, is used for low altitude, low latitude and GTO launches.

All the above sites are usually closed to foreigners, although the press was taken to the Xichang Launch Centre in 1988 when China was negotiating with US for approval of its launch business.

China's main mission control centre, Beijing Aerospace Command and Control Centre, is near Beijing where engineering labour colonies for assembly and testing of manned capsules are also located. Cosmonaut landings are on the vast Mongolian plain. They have also at least four, well-equipped tracking dishes in addition to Xian Satellite Control Centre and Namibia Tracking Station. They have established additional tracking stations in Karachi, Malindi and in Mongolia.

Launch Capabilities

China's first launcher, the Long March Rocket named Chang Zheng (CZ) developed with Russian assistance, was a three-stage, liquid-fuelled rocket which was used for launching their first satellite named Mao-1 weighing 170 kilogram into a near earth orbit in 1970. Since then China has developed a series of rockets using solid, liquid as well as cryogenic fuels for a variety of launches including manned missions. China has, at least, 12 types of launchers, of which the major ones currently in use and capable of launching up to 10-tons payloads into low earth orbit or 7 tons into GTO, are listed in Table 3.2. China is reported to be designing new generation of launch vehicles, which are non-toxic, pollution-free, low-cost and of high reliability, for launching payloads upto 27 tons in LEO and 14 tons into GTO, comparable to NASA's shuttle. The new heavy lift launchers are projected to be fully operational to meet China's future plan for setting up its own space laboratory in 2016 and manned space station in 2020.

Table 3.2

Major Rockets Developed by China

Sl. No.	Rocket	Year of Development	Type	Payload (in kg)	
				Low Earth	GTO
1.	CZ-1	1970 (Retired)	3 Stage Liquid	300	—
2.	CZ-2	1975	2 Stage Liquid	500	—
3.	CZ-3A	1983	3 Stage Liquid	2800	1400
4.	CZ-4A	1988	3 Stage Liquid	5000	2400
5.	CZ-2E	1990	3 Stage Liquid	9000	5000
6.	CZ-4C	1996	3 Stage Liquid	9200	5100
7.	CZ-2F	2010	3 Stage Liquid	10000	6000
8.	CZ-2F	2013?	3 Stage Liquid	27000	14000

Major Scientific and Application Satellites Programme in China

The China Aerospace Science and Technology Corporation (CASC) is a large state-owned enterprise that builds seven different series of satellites for carrying out scientific and application satellite programmes. These include:

1. Feng Yun Weather Satellites (FY Series)
2. Dong Fang Hong Communication Satellites (DF)
3. Ziyuan Remote Sensing Satellites (ZY)
4. Haiyang Oceanography Satellites (HY)
5. Shi Jian Scientific Exploration Satellites (SJ)
6. Beidou Navigation Satellites (BY)
7. Lunar and Interplanetary Missions (Change)

China has so far launched about 170 satellites. Since 1986 China has marketed over 40 commercial space launches for other countries. Even though China faced many failures earlier, since 1990 they have achieved better than 90 per cent success rate. China benefitted greatly, particularly after the Challenger failure (which was followed by the Arianne failure), by launching satellites for other nations, thereby earning over one billion dollars. Most of the foreign launches have taken place from the Xichang Launch Centre.

Chinese Weather Satellite: Feng Yun Series (FY)

Feng Yun means 'wind and cloud'. After a few polar-orbiting (FY-1) series of satellites, China started launching FY-2 series of geostationary weather satellites. The FY-3 series launched in 2004, 2006 and 2008 were the next generation of polar-orbiting Sun-synchronous weather satellites. These satellites designed to work for about three years in space, were meant for forecasting weather conditions and monitoring bad weather, particularly convective rainstorms, thunderstorms and hailstorms round the clock. They also monitor developing sandstorms as well as air quality to provide early warnings. China is one of the few countries having both polar orbiting and geo-meteorological satellites, which have now become part of the world-wide observation network.

Communications and Navigation Satellites: Dong Fang Hong Series (DFH)

China refers to its communications satellites as Dong Fang Hong (DFH), which means 'East is Red'. China's large communication satellites launched after 1990 carried C, Ku, Ka and L band transponders, for meeting the growing demand for education, commercial television broadcasts, stationary and mobile telecommunications as well as data, voice and video transmissions. Since initiating satellite education and TV broadcasting about 15 years ago, more than 50 million people seem to have acquired college or technical degrees.

Remote Sensing Satellites: Ziyuan Series (ZY)

China calls its remote sensing, earth resource satellites Ziyuan (ZY), which means 'Resource'. First in the series was the China–Brazil Earth Resources Satellite (CBERS-1 or ZY-1) launched in 1999 having a resolution of about

20 metres. It is believed that Brazil made part payment to China for these launches through a barter deal by selling sugar.

Unfortunately, even though Brazil had informally sounded us for cooperation in building remote sensing satellites following our success in IRS-1A and 1B having a spatial resolution of about 30 metres, we could not respond to the barter deal model. Following ZY-1, ZY-2 was launched in 2002, followed subsequently by ZY-3 and ZY-4 using the Long March-4B rocket from the Taiyuan launch centre, capable of taking imageries with a resolution of 1 metre or better, comparable to the later IRS satellites. Chinese Scientists, like Indian Space Scientists, have extensively used these satellites to survey national resources, monitor crop growth, detect environment pollution, assess crop yield and monitoring and management of drought and flood disasters. Remote Sensing has also been extensively used to evaluate project sites, city planning, surveying, cartography and also for defence purposes. China cooperated with Canada in the development of Synthetic Aperture Radar Remote Sensing and with the Surrey Space Centre in the development of small and micro satellites, including knowhow transfer and training. China has now launched comprehensive and multiple remote sensing satellites, each carrying a medium and very high resolution optical imaging instruments as well as high resolution Synthetic Aperture Radar (SAR), to serve both civilian as well as defence requirements.

In spite of the fact that we were far ahead in terms of remote sensing capabilities in the beginning, we have now lagged behind in the last few years, particularly in operational use of remote sensing in agriculture and operationalization of precision farming both of which are essential for enhancing our agricultural productivity without causing land degradation. Advance prediction of agricultural output, which is essential for national planning still continues to be totally dependent on the old system based on eye estimate. The FASAL programme developed by the space department for advance estimation of acreage and yield of major crops has not been operationalized. With the submission of the Professor Vaidynathan Report—I was a member of his group—we hope that the agricultural as well as the space department will join hands to benefit from the advances in remote sensing technology for obtaining accurate estimates of acreage and yield of all major agricultural crops, essential for meaningful economic planning of the nation.

Oceanography Satellites: Haiyang Series (HY)

China's Haiyang (means Ocean) HY-1 and HY-2 oceanographic micro-satellites have carried radar altimeters, microwave scatterometers, ocean colour scanners and multi-channel microwave radiometers to carry out real-time monitoring of oceans and coastal zones for biological resources, pollution monitoring and monitoring of estuaries, bays and navigation routes.

Scientific Exploration Satellites: Shi Jian Series (SJ)

Retrievable satellites have been used to conduct experiments in life sciences, space environment, space materials and new technologies. China refers to its science exploration satellites as Shi Jian (SJ) satellites. China is also carrying out research in high energy astronomy, atmospheric phenomena and micro-gravity. Chinese National Space Administration (CNSA) also carried out a joint satellite-based space mission with the European Space Agency (ESA) in 2004 for investigating the earth's magnetosphere. It consisted of two satellites, Double Star consisting of an Equatorial Satellite (TC-1) in earth's orbit, and a Polar Satellite (TC-2) for studying solar effects on Earth's environment, including detailed investigation of magnetic reconnection phenomena in the geomagnetic tail.

Beidou Navigation Satellite (BY)

China started using navigation satellites of other countries in 1980 for aircraft and ship navigation, geological disaster monitoring and forest fire detection. Since joining COSPAS-SARSAT, China's utilization of navigation satellites has remarkably increased and now covers even urban traffic control. China has already started setting up its own navigation satellite system with 25 satellites, which is expected to be completely operational before 2017.

Lunar and Interplanetary Missions

China launched its first Lunar Probe Change-1 (Legendary Chinese Goddess who flew to the Moon) into an orbit of 200 kilometres above the Moon in October 2007, to become the fifth country to orbit the Moon. They launched their second Lunar Mission Change-2 in October 2010 for carrying out scientific experiments, which were terminated eight months later, on 16 June 2011. Change-2 was then commanded to leave its Lunar Orbit and fly to the Lagrangian Point L2 to carry out sophisticated astrophysical studies. China now plans to carry out unmanned Lunar Landing and Rover Mission in 2012–13. It intends to accomplish a manned Moon-landing and bring back soil samples from the Moon in 2017, two to three years earlier than the US plan of man landing on the Moon.

Chinese long-term goal seems to be, to set up 'a base on the Moon and eventually mine its riches'. China is exploring the possibility of mining He^3 on the Moon and transporting it back to the earth for commercial energy production. It is known that thermo-nuclear energy from He^3 can be abundant and is very safe as the Deutrium-He^3 reaction produces less than 1 per cent neutrons, compared to the Deutrium-Tritium (D-T) reaction. D-He^3 reaction can be easily converted into electrical energy with more than 80 per cent efficiency. Unfortunately He^3 is practically non-existent on the earth, unlike on the Moon where He^3 is trapped by the mineral Ilmenite ($Fe\ Ti\ O_3$). Estimated amount of He^3 on the surface of the Moon is well over 2 million tons. Since 500 tons of He^3 is adequate to meet a year's energy requirement of the whole world, even the conservative estimate of He^3 on the Moon can supply adequate energy for the entire world for at least 4,000 years. China started developing its probe for its planned Mars mission in 2013, in collaboration with the Russian Space Agency. China is also reportedly planning to carry out a crewed mission to land on Mars during 2030–40.

Unlike China, India's space programme till recently concentrated on developing space technology and its expensive applications primarily to provide socio-economic benefits to the country. Even though India sent its first cosmonaut Rakesh Sharma on 3 April 1984 on a Soviet Mission (Soyuz T-11), India has started thinking of its own manned mission only recently. India successfully launched Chandrayaan-1 in 2009 to become the 6th country to do so and obtained significant scientific results, the most important of which was the discovery of water on the Moon. India is now planning to launch Chandrayaan-2 in 2013, which will carry a Russian Lander and an Indian Rover. India is also planning an unmanned mission to Mars in the next few years.

China's Manned Programme

China's manned programme was initiated around 1992, following the development of Long March CZ-2E capable of carrying a payload of over nine tons. The Shenzou space capsule (Shenzou means 'divine vehicle'), modelled after the Russian Soyuz spacecraft, was initially used to successfully orbit and recover monkeys, dogs, rabbits, etc. In 2000–01, Shenzou-3, circled the earth 108 times and then safely landed on the earth. In October 2003, China sent its first Taikonaut (derived from Taikong meaning 'space'), Yang Liwei, into space on Shenzou-5 from the Jiuquan Launch site in the Gobi desert in the Gansu province; he landed safely after spending 20 hours in outer space. This launch station in Gobi desert is used for all recoverable and microgravity missions and manned flights. Even in their very first manned flight, Yang touched down within three miles of the intended target. China followed that feat by launching two astronauts, Nei Haisheng and Fei Junlong, in its second manned mission on Shenzhou-6, which circled the earth for 5 days. China carried out its third manned flight in 2008 carrying three astronauts on Shenzhou-7. In this flight, astronaut Zhai Zhigang became the first Chinese to carry out space walk. Shenzhou, like Soyuz, has become a very successful work-horse "Space Transport Capsule", which is expected to play an important role in the realization of China's ambition to build a space station of its own by 2020.

A manned space station requires development of unmanned as well as manned docking module to ferry astronauts, supplies, etc., to the station. Accordingly, China has developed a three-step plan for accomplishing space rendezvous and docking. On 29 September 2011, China successfully accomplished its first step of launching an unmanned space laboratory module, Tiangong-1, using the Long March-2F rocket. The 8.0 ton Tiangong-1 (heavenly palace) which serves the crucial role as a docking target, has 15 cubic metre space where two or three Taikonaut's can easily live and work. Tiangong-1 can stay in space for at least 2 years. Shenzhou-8, an unmanned spacecraft was successfully docked with Tiangong-1 in orbit at 343 kilometres on 3 November 2011, which demonstrates China's mastery in carrying out such delicate manoeuvres. China has now offered its Long March 2-F rocket, which successfully launched Shenzhou-8, for launching heavy satellites for other countries.

An alternate to using heavy-lift launch vehicles is to continue with medium lift launchers, which have quicker turnaround time and cheaper, provided 'propellant depots' which can refuel spacecrafts in orbit could be set up in space, an option which is also being seriously considered by NASA. In the next two years, China intends to dock a manned module in space, following which it hopes to succeed in landing a Chinese Taikonaut on the Moon in 2017, at least three years ahead of the US which plans to return to Moon landing in 2020. China also hopes to establish a full-fledged observatory on the Moon in 2017. Robert Bigelow, a reputed space entrepreneur, believes China will not only land on the Moon in the next few years, but will also claim its own territory on the Moon, which could become a reality in spite of the UN's Outer Space Treaty, of which China is a signatory and which prohibits any such territorial claim.

Till recently, India's space programme had concentrated on developing its technology primarily for serving a wide range of applications towards providing socio-economic benefits to the country. These involved development of communication, remote sensing and meteorological satellites for providing nation-wide broadcasting, communication, tele-education, meteorological observations, disaster mitigation and efficient management of agriculture, water, mineral, forest, marine and environmental resources. It is only whenever a window of opportunity was available, scientific experiments in aeronomy and astronomy were accommodated as piggy-backs on some satellites. Only recently, satellites such as Chandrayaan-1 or ASTROSAT, scheduled for 2013, and Chandrayaan-2, also scheduled for 2013, which are dedicated for carrying out scientific explorations became a part of ISRO's programme. In addition to exploring the possibility of carrying out a Mars mission, India is also in the process of developing a modest medium-lift launch vehicle with an advanced cryogenic engine and studying the possibility of including manned mission as a part of its future programme.

Development of Ballistic Missiles

Even though China built and launched its first missile with a range of about 500 kilometres in 1958, it started to develop its DF-5 intercontinental ballistic missile in 1964, just about a year after it carried out its first nuclear test in 1964. DF-5 was designed as an intercontinental ballistic missile (ICBM) with a range of 12,000 kilometres. China has now developed a series of both nuclear and non-nuclear missiles of all types ranging from inter-continental with a range of over 12,000 kilometres to short range, surface as well as submarine launched missiles. According to estimates, China has a stock of over 100 inter-continental ballistic missiles and adequate MIRV warheads. Table 3.3 is a brief summary of the Chinese Missiles presently in use. Figure 3.2 shows the strike contour capabilities of China's major nuclear and conventional missiles (Figures 3.1, 3.2 and 3.3 provided at the end of this chapter).

Table 3.3

Major Chinese Ballistic Missiles in Service

Type		Missile	Type	Range in Km
(A)	Short Range DF-11	CSS-6	Single Stage SRBM	500
	DF-15	CSS-7	Single Stage SRBM	800
(B)	Medium Range DF-21	CSS-5	2 Stage MRBM	1770
	DF-21C, 21D	CSS-5	2 Stage MRBM	1750
(C)	Intermediate Range DF-3	CSS-2	Single Stage MRBM	3000
	DF-4	CSS-3	2 Stage IRBM	5000
(D)	Intercontinental DongFeng-5A	CSS-4	2 Stage ICBM	11000
	DF-31	CSS-9	3 Stage ICBM	11700
	DF-31A (DF-41)	CSS-9	3 Stage ICBM	13000

Type	Missile	Type	Range in Km
(E) Land Attack Cruise DongFeng-10	DH-10	Single Stage MRBM	3000
(F) Submarine Launched Julang-1 Julang-1A	CSS-N2 CSS-N3	2 Stage 2 stage	1770 2500
Julang-2	CSS-NX4	3 Stage, ICBM	8000

Indian defence initiated the Integrated Guided Missile Development Programme (IGMDP) in the early 1980s for the development of a comprehensive range of missiles including intermediate-range Agni missile (surface-to-surface) and short-range missiles such as Prithvi (surface-to-surface), Akash (surface-to-air), Astra (air-to-air), Trishul (surface-to-air) and Nag (air-to-air), Nag (anti-tank). Prithvi, being liquid fuelled, is highly accurate. Agni missiles have a range varying from Medium (MRBM) to Intermediate (IRBM) and Intercontinental (ICBM) ranges. The recently-launched Agni has a range of 5,000 kilometres. In addition to Sagarika, India has also developed a two stage surface-to-surface submarine launched missile, a variant of Prithvi and Dhanush with a range of 700 kilometres. A ship-launched variant of Prithvi with a range of 350 kilometres has also been test-fired.

Brahmos missiles, named after Brahmaputra and Moskova rivers, a joint project of India and Russia which can be land-, ship- or air-launched are highly accurate tactical missiles, ready for induction into the Indian Army, Navy or Air Force. Sagarika is a nuclear-capable submarine launched ballistic missile with a range of 750 kilometres. India has also developed Pradhyumna Ballistic Missile intercept, which can intersect/ engage ballistic missiles of 300 to 2,000 kilometres class at a speed of Mach-5. The Defence Research and Development Organisation (DRDO) is also working on missiles intercepting targets of 5,000+ kilometres range at an altitude of 150 kilometres. Likewise Akash, a medium-range surface to air missile with an intercept range of 30 kilometres and propelled by solid fuel, as well as Akash surface-to-air missile supported by multi-target and multi-function phased array radar have also been developed. The other missiles developed by DRDO include Trishul and Nag, short range surface-to-air and surface-to-surface missiles. Table 3.4 lists the major missiles developed by DRDO.

Table 3.4

Major Indian Missiles in Use

Missile	Type	Warhead	Fuel / Stages	Range (km)	Payload (kg)	In service
Prithvi-I	Tactical	Nuclear, HE	Single stage liquid	150	1000	1988
Prithvi-II	Tactical	Nuclear, HE	Single stage liquid	350	350–750	1996
Prithvi-III	Tactical	Nuclear, HE	Single stage solid	350–600	500–1000	2004
Agni-I	Strategic	Nuclear, HE	Single stage solid	700–800	1000	2002
Agni-II	Strategic	Nuclear, HE	Two and half stage solid	2000–3000	750–1000	1999
Agni-III	Strategic	Nuclear, HE	Two stage solid	3500–5000	2000–2500	2011
Agni-IV	Strategic	Nuclear, HE	Two stage solid	3500	1000	Test fired
Agni-V	Strategic	Nuclear, HE	Three stage solid	5000	MIRV	Test fired 2012
Brahmos	Cruise	Submarine, land, air launched	Two Stage	300	300	Tested fired 2010
Sagarika	Tactical	Submarine launched	Prithvi-2 based	700	1000	Tested – 2005
Shourya	Tactical	Submarine launched Nuclear, HE	Two stage Solid	700	—	Tested-2010
Dhanush	Tactical	Submarine launched	Prithvi-2 based	350	1000	Test fired 2005

Reconnaissance Satellites

Yaogan-1, the first Chinese reconnaissance satellite weighing 2,700 kilograms and carrying both electro-optical sensors and L-band SAR was launched in 2006 from Taiyuan Launch Center. It is believed that Recon satellites are owned and operated by the People's Liberation Army (PLA), unlike the Civilian Remote Sensing Satellites. Yaogan-2, launched from Jiuquan Center, had CCD cameras with a resolution of 1.5 metres, which were followed by Yaogan-3, 4 and 5. Yaogan-6, launched from Taiyuan, carried a second generation SAR, which was repeated in 2010. The list of Reconnaissance Satellites launched so far by China is given in Table 3.5.

The Defence Research and Development Organisation of India is reported to be developing an Advanced Reconnaissance Spy Satellite (CCI-SAT) likely to be launched in 2014, into a 500 kilometre-circular orbit. This satellite is supposed to be capable of taking high-resolution images using SAR. Compared to recently-launched RISAT of ISRO, CCI-SAT is claimed to be more sensitive in addition to it having a provision to steer along/across the track.

Table 3.5

Reconnaissance Satellites Launched

Name	Date	Launcher	Launch Site	Orbit
Yaogan-1	27 Apr 06	CZ-4C	Taiyuan	SSO
Yaogan-2	25 May 07	CZ-2D	Jiuquan	LEO
Yaogan-3	12 Nov 07	CZ-4C	Taiyuan	SSO
Yaogan-4	1 Dec 08	CZ-2D	Jiuquan	LEO
Yaogan-5	15 Dec 08	CZ-4B	Taiyuan	SSO
Yaogan-6	22 Apr 09	CZ-2C	Taiyuan	SSO
Yaogan-7	9 Dec 09	CZ-2D	Jiuquan	LEO
Yaogan-8	16 Dec 09	CZ-4C	Taiyuan	SSO
Yaogan-9	5 Mar 10	CZ-4C	Jiuquan	LEO
Yaogan-10	10 Aug 10	CZ-4C	Taiyuan	SSO
Yaogan-11	22 Sept 10	CZ-2D	Jiuquan	LEO

Note: SSO: Sun Synchronous Orbit, LEO: Low Earth Orbit.

ASAT Weapon Development

Disabling of enemy satellites using weapons such as war-head detonators, electromagnetic weapons, ground-based high powered lasers, kinetic killers, nuclear weapons etc., have been experimented since the late 1960s by Russia and the US. While anti-satellite (ASAT) testing by the superpowers was practically stopped during the last decade, China tested its first anti-satellite kinetic-kill weapon, boosted on a two-stage DF-21 launcher to successfully destroy an old Chinese weather satellite in the 865 kilometre polar orbit in January 2007. In this process, thousands of space debris fragments were dumped between 800 and 1,000 kilometres altitude in the atmosphere, which will stay for decades in space. Following this incident, the US used a direct ascent kinetic energy interceptor in 2008, to successfully destroy an US failed spy satellite, which was about to re-enter the atmosphere. The US and Chinese experiments prompted Russia to restart its own ASAT programme which was stopped in 1993. In spite of many countries supporting a ban on ASAT weapons, there has been no progress.

China has developed destructive energy weapons involving high powered lasers and microwaves, which were tested during 2006–08. They also developed micro-satellites of 10–300 kilogram-class, which could latch on to any targeted satellite and destroy it. China has successfully demonstrated its capability to destroy high altitude ballistic missiles. Recently, DRDO has claimed that it is developing, on a priority basis, high-energy lasers and exo-atmospheric kill vehicles to destroy enemy satellites.

Conclusion

China's Goals

Short Term Goals in Development Target

1. Build comprehensive earth observation system involving meteorological, resource, oceanic and disaster monitoring satellites for dynamic monitoring of land, atmosphere and oceans.

2. Set up robust satellite broadcasting and communication system. Some of the telecommunication satellites launched by China, such as Feng Huo-2 and Sinosat-2 in 2006 are reportedly controlled by the People's Liberation Army. In addition to successful development of ASAT weapons, China has achieved capability in carrying out information warfare and attack enemy computer networks using viruses.

3. Establish an independent satellite navigation and positioning system.

4. Upgrade overall capacity of China's launch vehicle, which must be low cost, non-polluting and high performance (70 tons at low earth orbit?).

5. Intensify manned space flight programme and establish advanced manned space projects.

6. Intensify space science activity and carry out intense activity in outer space and planetary exploration including on Moon and Mars.

Long Term Objectives

1. Establish a strong and robust manned space flight programme.

2. Establish superiority in the exploration and utilization of space science and technology.

3. Establish multi-function and multi-orbit infrastructure in space including space station platforms and space habitats.

4. Achieve world leadership in industrialization and marketization of space technology and its applications.

5. Build full range of space capabilities in intercontinental ballistic missiles, ASAT weaponization, space-docking and space exploitation.

In summary, China's space programme is aimed at achieving a full range of space capabilities including a comprehensive science and application programme, development of powerful launch vehicles, space-docking capabilities, short-range tactical as well as long-range intercontinental ballistic missiles, ASAT weapons and manned flights. In the next 10 years, China is planning to land Chinese Taikonauts on the Moon, possibly followed by establishment of manned flights to Mars in the next 20 years. It has built a vast space infrastructure including deep space tracking, telemetry and radar systems. Knowledgeable estimates indicate that China has launched over 170 satellites, about 40 of them for other countries on a commercial basis (as against just five over 100 kg by ISRO) and earned over $1 billion (as against just $60 million by ISRO). Chinese defence industries since 1990 have evolved from essentially being government ministries to organizations that can effectively operate in the global market. China has started setting up its own GPS system called Beidou based on 25 satellites, which is expected to become fully operational in the next five years (around 2017).

As of now even the recently launched Agni-5 with a range of about 5,000 kilometres is no match to Chinese intercontinental ballistic missiles.

China, being well aware of the implications of American policy on space as spelt-out by the Rumsfeld Commission, has articulated its own policy almost along the same lines. Like the US, China intends to achieve a full spectrum dominance, which is defined as:

1. Control of space to assure or deny access to and freedom to operate in space.
2. Global engagement including establishment of surveillance, command, control and communication functions.
3. Full force integration requiring total integration of space forces with land, sea and air forces and
4. Global partnership—augmenting military and space capabilities through exploitation of civil, commercial and international space capabilities or through bilateral partnership.

China has fashioned its strike capability to be able to effectively counter US superiority in space arena. Outwardly, China, like India, has advocated space treaty for preventing an arms race in outer space including ban on placing weapons in space orbit. In reality however, China's military posture, its mastery and level of space technology development, ambition to establish itself as the most powerful space power and economic advancement, which enables it to spend huge amounts of money on development of space technology, not only threatens the US superiority and its capability but also has serious implications for our own country.

Over the past decade, Chinese defence has totally transformed from an infantry-dominated force with limited power projection ability into a modern force with Long March precision strike assets. The estimated strike capability of China includes:

1. Over 1500 short-range, precision guided as well as intercontinental ballistic missiles supported by well-equipped air power and electronic warfare.
2. Adequate fire power to destroy enemy capabilities on enemy territory, to back-up its declared policy of pre-emption as logical posture for confronting even a powerful enemy.
3. Full support to achieve revolution in military affairs, through asymmetric warfare and informatized weapons. In China's texts, effective engagement involves not only demonstration of military strength, but also a comprehensive contest on all fronts such as politics, economy, diplomacy and law. Heavy investments in space are an integral part of any asymmetric warfare.
4. Open declaration to become a world leader in the development and use of space technology and achieve the capability to orbit any type of satellite 'within 24 hours upon request'.

In summary, Chinese defence strategy consists of:

1. Stock up adequate number of medium, long range and intercontinental missiles with very high yielding and powerful conventional as well as nuclear warheads.
2. Shift its nuclear doctrine from 'no first use' to 'use against any nuclear equipped power, which shows its intention to attack China'.
3. Projection of China's naval power in the Indian Ocean. In addition to the two naval ships patrolling the Somalian Coast, China is also reported to be exploring establishment of naval/military bases in littoral countries of Indian Ocean.
4. Establishment of strong synergy between China's civilian and military programme.
5. Clear recognition of space having emerged as the most powerful fourth medium of war.
6. Capability to respond to any external threat within 24 hours.

Estimated nuclear warheads in China are about 3,000 (the US and Russia have over 6,000 each). China has a fleet of about 1,500 missiles (to which 100 are added each year) including tactical and nuclear. Nuclear missiles are stored underground (like in North Dakota in the US or Novosibirsk in Russia) and inside tunnel complexes (estimated tunnel network is 3,000 miles long).

Budgetary Summary

The basic difference between India and China arises due to China's substantially higher GDP and hence its ability to spend much more on defence including space. China's GDP is estimated at about $4 trillion as against $1.5 trillion of India or 15 trillion of US. China's defence budget, including defence-related space budget in 2010, is officially stated to be $45 billion, even though independent intelligent estimates vary between $90 to $120 billion. Figure 3.3 shows the official budgetary allocation to Defence and Space in China since 1994 along with estimates from other independent sources.

Conclusion

1. India cannot and should not overreach its financial resources just to compete with China in Space. Nor can we afford with the multitude of problems we face such as lower GDP, large-scale poverty, poor infrastructure, industrial backwardness and, above all, rampant corruption which has dangerously gripped the nation.
2. Even though most of the satellites are dual-use satellites by nature which can be used for both civilian and defence purposes, the actual use of our civilian satellites by our military establishment is very poor.
3. India's space programme needs to be carefully tuned to forge a better interaction with our defence. The development of critical technologies, like heavy-lift launch vehicles, more purposeful and well-designed satellite systems are important to ensure achievement of maximum benefit at least cost.
4. While china has long been an advocate of space arms control, unlike the US, China's ASAT testing in 2007 has raised doubts

about its real intention. Some experts feel that bringing pressure on the US and making the US to abandon its ASAT programme, is their objective with a view to buying time and catching up with the US.

5. India needs to carefully evaluate and selectively build its own capability in dual-use technology. We need to clearly evaluate how far we should go in terms of building more powerful rockets, other forms of space transportation systems, human presence in space etc., based on clear cut goals and affordability. This requires constitution of a high-powered think-tank, which should ensure better coordination and collaboration between civilian space and defence sectors in planning and developing dual use systems.

6. Due to large investments and high visibility of their Space Programme, the Chinese have been able to attract the best young minds in large numbers. China firmly believes that investments in dual-use technology are very important because of the high rate of return on investment.

7. India has to carefully tailor its own military strategy and diplomacy in the Asian Region, taking into account China's capability instead of being completely Pakistan-centric.

8. It is extremely important to urgently constitute a small team of experts from defence and space to chalk out a cohesive programme that can advance the goals of both organizations.

Whereas corruption-free administration is an essential requirement for the country to progress, it alone is not sufficient. Technical capability must be significantly enhanced if we are to effectively tackle the powerful and well-equipped enemies. With declared policy of no-first-use of nuclear arsenal, we must be able to withstand first and second attacks and be able to quickly inflict severe massive retaliatory damage on the enemy. It is also clear that effective and lethal counter attack is possible only when critical elements of the attack are appropriately located at scattered locations but could be assembled within the shortest time possible to mount an effective counter attack.

There is a tendency to over-claim and over-advertise our accomplishments, which without producing any effect on the knowledgeable enemy, is creating a false level of comfort within the country. If we truly want to build our capability, we have to tap the vast potential in terms of technical capability and infrastructure built in the Indian industry, like the US, Russia and China. A greater synergy must be established between our defence establishments and civilian establishments of space and nuclear energy and ensure the full intellectual strength of our country is employed in planning and execution of our national strategy.

Figure 3.1

Maximum Range of China's Conventional SRBM Force

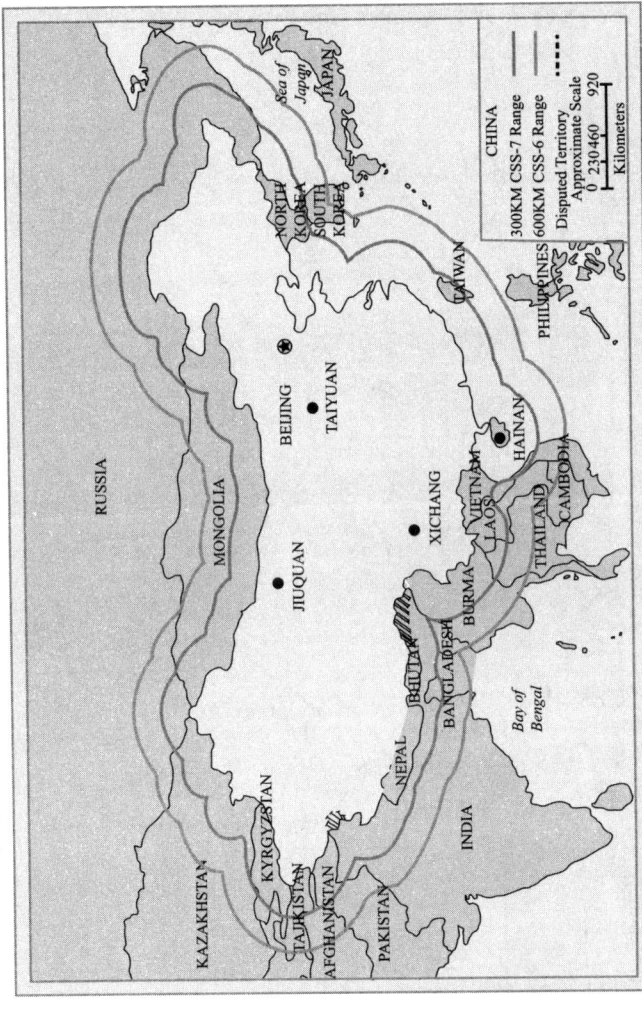

Source: Annual Report on Military and Security Developments Involving People's Republic of China—Office of the Secretary of Defense (2007), 23.

Figure 3.2

Strike Contours of Medium and Intercontinental Range Ballistic Missiles

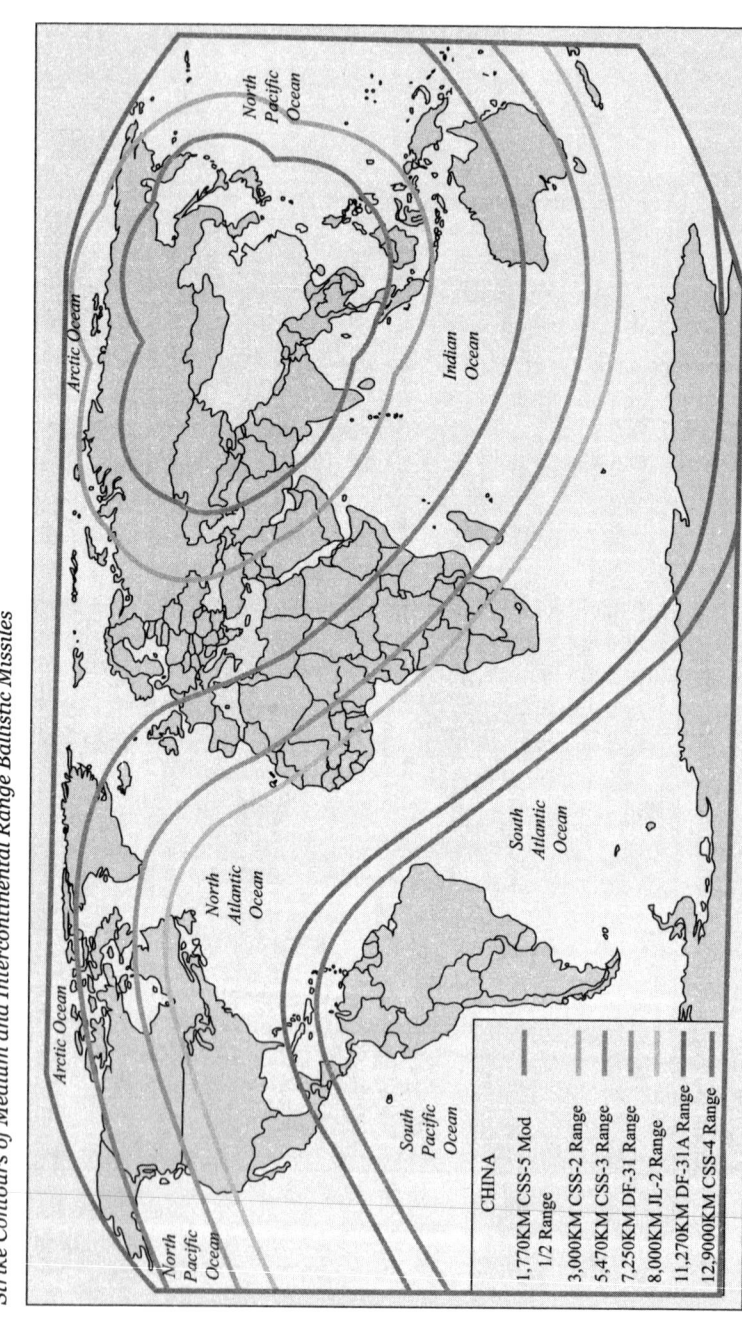

Source: Annual Report on Military and Security Developments Involving People's Republic of China—Office of the Secretary of Defense (2007), 19.

Figure 3.3

Chinese Defence Budget and Estimates of Total Defence-related Expenditure

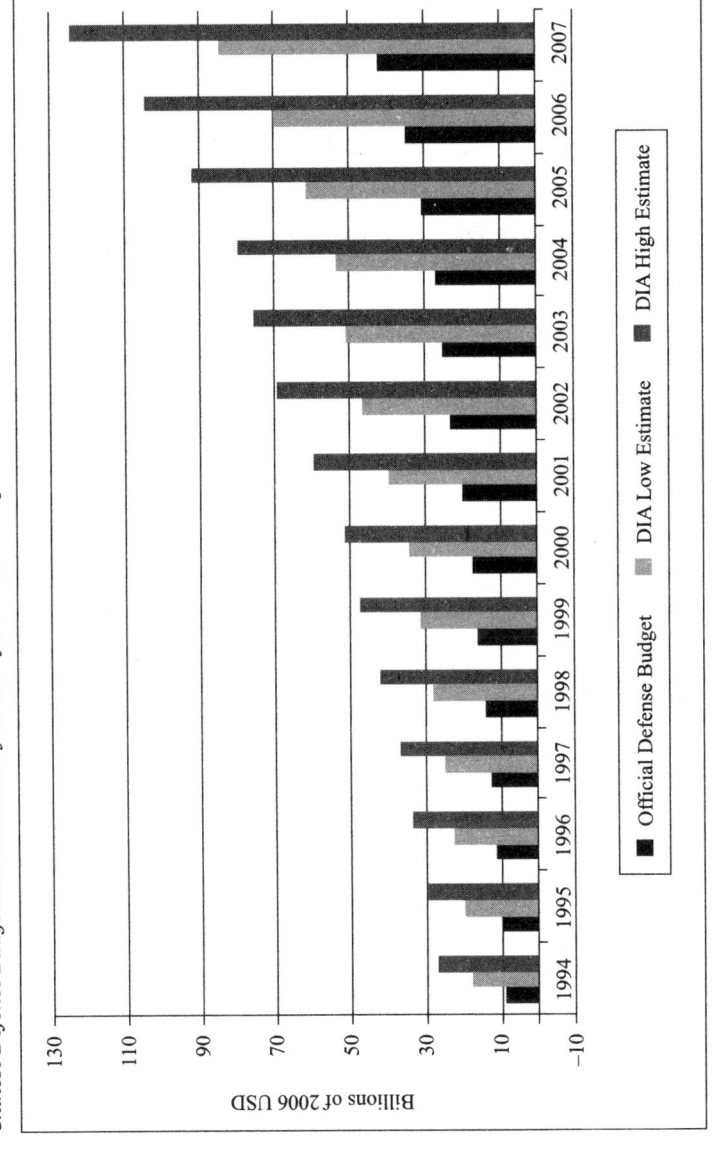

Source: Annual Report on Military and Security Developments Involving People's Republic of China—Office of the Secretary of Defense (2007), 27.

4

The Flying Dragon

*Is China Set to Emerge as a Global Player in Aeronautics?**

Roddam Narasimha

*Since Deng Xiaoping declared some three decades ago that 'to get rich is glorious',
the Chinese mantra appears to have been: 'get rich, export and do S&T (roughly
in that order), and everything else will follow.'*

Introduction

The Chinese aeronautical ecosystem has developed rapidly in the last
10 to 15 years and, if recent progress can be sustained, is set to emerge
as a major global force before long. This development is of particular
importance to India for a variety of reasons, including both strategic and
commercial. Smita Purushottam mentions aerospace in her contribution:
I shall here confine myself to the aero component.

China's policy today in aviation and aeronautical industry can be
summarized as driven by global ambitions, in both civil and military
sectors. From the Large Civil Aircraft (their LCA) to a stealth aircraft
which flew for the first time about a year ago (well ahead of western

---

*   Much of the material in this chapter is taken from the technical press and
    various websites. In order not to clutter the article, citations have been given
    only at selected places in the text.

experts' estimates), China is challenging current global leaders virtually across the board. Their objective is clearly to be an aerospace superpower, and they are pouring in huge resources—both human and material (a quarter of a trillion US dollars [nearly ₹ 15 lakh crores], according to *The Economist*)—into the project. It appears to be part of their pursuit of what they call comprehensive national power (CNP). I shall conclude with an analysis of that concept.

In the early Mao years, China primarily sought a deterrent against the United States, and the national project emphasized missiles, space vehicles and nuclear weapons. Aircraft were mostly imported from the Soviet Union. The situation today is totally different: the C919, intended to be a competitor to the Boeing 737 and Airbus A320, now appears to have about the same national status as the manned space-flight programme. A recent report in *Aviation Week & Space Technology* says that China is now a nation gone crazy about making aircraft.

Civil Aeronautics

The Chinese civilian programme, a major initiative that was announced in the 2007 National People's Congress, gives priority to commercial over military aviation. At first glance this seems surprising, but it appears consistent with the Chinese concept of CNP, in which economics (especially, foreign trade), science and technology are given dominant weightage. This civilian initiative comes with the support of China's Commission on Science, Technology and Industry for National Defence (COSTIND),[1] which prepares national plans in all sectors of S&T and industry, including commercial aviation.

In July 2012, the Chinese State Council declared that its development target would be to promote [a] 'safe, convenient and highly efficient' ... 'internationally competitive [aeronautical] industry by 2020', and to provide aviation services that would cover 89 per cent of the population of China. Among the basic principles adopted by the State Council was the coordination of civil aviation and the 'military route'.

The Chinese civil initiative is partly driven by the realization that the centre of gravity of global civil aviation has now shifted to Asia. Boeing estimates that between now and 2030 demand for new civil aircraft—

from regional to large aircraft (400+ seats)—will be 33,500, of which 11,450 will be from India-Pacific, i.e. about a third (whether by value [\$ 4.5T] or by number, it turns out): China alone will need some 5,000 (more than thrice the demand from South Asia).

Following the 2007 decision, major structural reforms were instituted in 2008. These included the establishment of two new corporate entities: the Aviation Industry Corporation of China (AVIC) and the Commercial Aircraft Corporation of China (COMAC). AVIC traces its history to an Aviation Industry Bureau set up in 1951 during the Korean War; it went through a series of some 10 reforms through the years before it took its present form. It is a state-owned aerospace and defence company (or rather consortium), with more than 10 major affiliate companies and more than 400,000 employees. (For comparison, the total number of employees in HAL, which is the only significant aircraft industry in India, is probably less than 40,000.) COMAC is a subsidiary of AVIC, in which investment has been made by the State-owned Assets Supervision and Administration Commission of the State Council, by AVIC itself, and by a variety of other agencies including the Shanghai Municipal Corporation.

Shanghai won its investment share against competition from several other cities and provinces chiefly because of its strong international contacts. As COMAC linked 10 industries together, it became a huge enterprise, with a capitalization of US\$ 2.7 billion (approximately ₹ 15,000 crores). Explicitly set up with a mandate for implementing large civil aircraft programmes in China (including, in particular, the C919), COMAC holds a special position in the country as it reports directly to the State Council, not to AVIC. It has chosen to adopt the principle of 'development with Chinese characteristics'. According to its website, COMAC attaches great importance to 'technological progress and self-reliant advancement in such areas as marketing, integration, localization and globalization' [!], and is 'determined to independently build large Chinese passenger aircraft that will soon be soaring through the blue skies'. (The word 'self-reliant' sounds old-fashioned in India today.) Furthermore, the C919 programme is 'the will of China and her people', and will be able to claim indigenous intellectual property. It will furthermore 'bring into full play the political supremacy of the socialist system which is capable of concentrating its resources in achieving great things'. In a message, the Chairman of COMAC explains how large commercial aircraft development promotes S&T progress in China and its comprehensive national power (the last word appears as 'strengths' in the original). The message is a display of what the Chairman calls 'aeronautical patriotism'. COMAC's 10 affiliate industries come from across China, including those in Shanghai, Xian, Harbin and Chengdu. The scale on which the Chinese act on their plans is indeed impressive.

At the end of this chapter are some pictures of Chinese aircraft. Some of these are new aircraft that have either just entered service or will do so in the very near future. The MA700 (capacity 70–90 seats?) is a new turboprop. An earlier turboprop, the MA60 (capacity around 50 seats), was not an extraordinary success: some were sold, some given away (donated) to friendly countries such as Myanmar and others in Africa and central Asia. The MA700 is being designed by AVIC, and is expected to enter service by 2017. It will compete with the two major aircraft in the same class, the ATR-72 and the Bombardier Q400, both now flying in India. The MA700 has a wider and more spacious cabin than either of the other two, but is shorter and lighter. Incidentally, China now supplies major fuselage sections as contractor to Bombardier on its 110–149 seat jet family CS100, but competes with it on the turboprop Q 400. The Chinese government is investing about US$ 400 million of public money in Chinese industry for work on the Bombardier C-series aircraft. The need for this is perhaps highlighted by reports that the Chinese partner has been having some serious manufacturing problems, with Bombardier temporarily shifting production from China to avoid delays in the first flight of the CS 100. According to 11 April Wall Street Journal, the first flight is now expected sometime before June 2013 and first delivery will be around mid-2014, but a delay now seems likely.

Then there is the advanced regional jet ARJ21 (now a project with COMAC), which can carry 90–100 passengers at a normal cruise speed of Mach 0.78. (Civil transport aircraft speeds have slowly declined in recent years in response to the higher price of oil and the demand for lower green-house gas emissions due to climate-change concerns.) The aircraft will be powered by General Electric CF34-10A engines from the US. It is thought to resemble the Douglas DC-9 family, and has received assistance from the Russian Antonov agency; it is yet to be certified.

The most ambitious civil aircraft project is the COMAC C919 currently under development—a 'trunk-liner program [that] is the will of China and her people'. (In the name, C stands for China or COMAC, 9 is a numerical symbol which means 'Forever' in Chinese, and 19 is for 190 seats.) This was approved in 2007 as part of the eleventh Five-Year Plan. (Indian and Chinese five-year plans are in near synchronism.) Its objective is to challenge the A320 and the Boeing 737, which (in terms of total number of aircraft sold or selling) are the most successful commercial aircraft ever made in the world. The 737 made its first flight 45 years ago and is close to selling its 10,000th unit in the line. The A320 was designed much more recently (first flight 25 years ago) and with a higher technology than was available on the earlier Boeing 737s—such as fly-by-wire controls. The two aircraft are in strong competition with each other: it will be interesting to see if the C919 joins them (making global civil aeronautics dominated by what is being called ABC). Interestingly, Airbus has set up a plant in Tianjin for the manufacture of the A320; since 2009 the Tianjin plant has delivered as many as 89 A320s - in spite of higher costs in China than in Europe. Even as the negotiations were going on with Airbus regarding the manufacture of the A320, China was planning its own challenge to the same A320 in the form of the rather similar COMAC C919.

The C919 is scheduled to enter service as early as 2016. It is designed around imported engines, avionics, and various other systems. It has a cruise speed of Mach 0.785, a little lower (again to save on fuel) than what the A320 would normally do, and will be made in two versions. They have a launch order of 100 aircraft from Chinese airlines, and expect to sell 2,000 by 2030.

In embarking on a project like C919, the Chinese have benefitted from the international supply-chain/outsourcing policy of Boeing on the 787 'Dreamliner'. This policy resulted in substantial orders for 787 components on Chinese companies. The Chinese now follow the Boeing philosophy. Although many C919 components or subsystems may be manufactured within China, many others would be outsourced. Thus the electrical systems will come from Sundstrand and auxiliary units from Honeywell and Goodrich. Indeed major western aircraft subsystem vendors are beating a path to AVIC, seeking a share of what they see as a potentially huge market. COMAC has thus ended up with 13 Western suppliers as partners. However many western companies working in China are in something of a dilemma: on the one hand they fear that their technologies may be copied, on the other they cannot afford not to work with the Chinese because of the huge market. So Chinese strategies seem to be working!

While on the subject, this dilemma faced by many western companies is illustrated also by the case of Ameco, China's largest maintenance service provider. Ameco is 60 per cent owned by Air China and 40 per cent by Lufthansa Technik. So is Lufthansa teaching a potential rival how to compete? Western companies are both lured and scared at the same time—lured by the potentially huge profits from sales, scared by the potentially large losses in IP for China's record in respecting IP is poor, in contrast for example to Embraer, which is both IP-honest and profitable.

China's goals in aerospace are, at one and the same time, ambitious in vision and pragmatic in execution. For example, what strategy should a new player in the global scene (like China) adopt to have its own aircraft penetrate a highly competitive commercial international market? Such an aircraft would have to offer appropriate customer service, possess passenger appeal, organize airline financing arrangements and so on. To train its employees on these issues, the aircraft manufacturer AVIC bought up Joy Airlines so that AVIC design engineers could gain first-hand experience with operational problems in an airline. So Joy operates AVIC aircraft like the MA60 and, based on their operating experience, AVIC engineers introduce necessary design changes in new aircraft they propose to launch (such as the MA 700). AVIC is also getting Lockheed Martin and Boeing to set up their celebrated 'smart' groups (Skunk Works in the former, Phantom Works in the latter) of clever designers with direct access to manufacturing facilities, enabling them to design and manufacture novel, innovative aircraft with unusual performance parameters. (One well-known example is the U2 [made by Skunk Works, led at that time by the legendary Kelly Johnson], which became famous during Cold War-days for its high-altitude surveillance flights over the Soviet Union.)

Similarly, Hawker-Beechcraft, a company with an Anglo-American legacy known for its small aircraft (including the King Air used for intelligence and surveillance operations), agreed earlier in 2012 (after having filed for bankruptcy in May) to sell part of its plant to a Chinese manufacturer for about $ 1.8 billion. This proposed deal has the US in another dilemma: Are the security concerns associated with selling the company to the Chinese weightier than the local political need to save jobs at Wichita, Kansas?

The current global civil aviation market is dominated by three 'duopolies': Airbus and Boeing in the larger narrow and wide-body classes (150+ seats), Embraer (Brazil) and Bombardier (Canada) in the large regional jet class (less than 150 seats), and ATR (Europe) and Bombardier in the larger turboprop (more than 50 seats) class. All these duopolies are being challenged: the first by China (C919) and Russia (UAC MS-21) (considered near-peers now in the West), the second by Mitsubishi (MRJ), Sukhoi (Superjet 100) and China, and the third only by China till now.

Finally, half of the global profits in civil aviation are being made by China, according to *The Economist*. It is no wonder, therefore, that civil aviation infrastructure in China is growing at a dizzy pace: in 2011, 34 major airports were under construction, 63 more are planned by 2020. The Civil Aviation Administration of China estimates that per capita air travel will increase fivefold in the next two decades (*Aviation Week and Space Technology [AWST]*, 28 February 2011). Similarly China will account for 10 per cent of the global market in 20 years.

Military Aircraft

Among the first projects in China's military aircraft industry was the F7, a version of the Soviet MiG 21, whose development started in the 1950s and 1960s. The project saw several ups and downs mirroring the state of Sino-Soviet relations, and full-scale production began only in the 1980s. An export version of the F7 with western systems was in service with the Pakistan Air Force.

In the late 1980s and 1990s, rapid economic growth enabled large investments in S&T and infrastructure, and China's policies began to change. In the last decade they have abandoned old models, embarked on a major military modernization plan, and adopted bold and even gigantic plans that have given a big boost to Chinese military aircraft design and development. We shall consider briefly some of these projects.

The most notable among them is the J10 military fighter. Its development has gone through several phases, achieving first flight in 1996 (eight years after the project started) using a Russian turbo fan (other Russian technologies and 'scientific guidance' also appear to have helped). The resemblance of its delta-canard configuration to that of the cancelled Israeli fighter Lavi has led to speculation about possible involvement of Israel. It is reported to use carbon-composite materials and has a digital quadruplex flight control system for its unstable configuration. It has been made in five variants including one that is sometimes called Super-10. As of February 2012, the People's Liberation Army has been estimated to have 260 units. There have been four reported crashes of the J10, including that of a prototype in 1998 attributed to the flight control system. The aircraft achieves Mach 2.2 at altitude and Mach 1.2 at sea level, and has a maximum take-off weight of 19.3 tonnes. With the acquisition of the Sukhoi-30MK the total number of J10 produced may be smaller than earlier planned.

China's stealth aircraft, the J20, had its first flight in January 2011 (to the surprise of Western observers, most of whom had thought it would take longer). Its design reveals the strong influence of the well-known U.S. aircraft F22 (in service), and the Joint Strike Fighter F35 (now in an advanced stage of development). (The Russian project MiG 1.42, which is itself thought to have been inspired by US work, might also have been relevant.) The J20 has a maximum speed of Mach 2.5 and a combat range of 2,000 kilometres. Not much is known about the aircraft in any detail: not even whether it is just a technology demonstrator or a prototype eventually to be productionized. US experts admit they cannot accurately assess its performance, but consider that the J20 may be vulnerable to new US radars. Stealth is usually achieved from aerodynamic shaping inspired by the demands of the electromagnetic characteristics of the aircraft required to achieve low radar cross-section, shielding jet engine exhausts and the use of radar-absorbing surface coatings. Today's stealth aircraft rely less on coatings than they did earlier, following problems with durability and the ageing properties of the coating materials; newer designs rely instead chiefly on aerodynamics and electromagnetics. Some US analysts believe that the US will, therefore, remain well ahead of China for another decade or two in stealth technology, and it remains to be seen how successful the J-20 will prove to be. A problem is that technology changes on a 10-year time scale, whereas the airframe lasts 30–40 years, so it is not clear how stealth technology can in practice be continually updated (*AWST*, 6 August 2012). This may in part be responsible for the delays and over-runs in the F35, the planned successor to the F22 as its production ceases in the US.

The Chengdu FC-1, also often called CAC/PAC JF-17 Thunder indicating that it is a joint project between the Chengdu Aircraft Corporation and the Pakistan Aeronautical Complex, is a multi-role combat aircraft. (The geo-political history of this project is murky, and the interested reader can look up reports in www.defenseindustrydaily.com.) It is powered by a Russian Klimov RD-93 or Chinese WS-13 afterburning turbofan. Its maximum speed is Mach 1.6. It is intended to be the back bone of the Pakistan Air Force, which plans to induct a total of 250 units. The first Pakistan squadron was set up in February 2010. In June 1995 Mikoyan of Russia joined the project to provide design support to Chengdu. From available accounts of the JF-17 and the announced plans for further development, it appears that in the first block the aircraft had little carbon fiber technology. It is also reported to have a 'computerized flight control system', probably a fly-by-wire system rather than one for an unstable aircraft. The second PAF squadron of JF17 was being equipped as of April 2011. Pakistan Air Force seem happy with the Chinese aircraft. Pakistan is also acquiring four ZDK-03 airborne early warning aircraft and the armed UAV CH-3 from China.

China also seems to have plans for combat drones. Western experts suggest it is about a decade behind.

However there is evidence that the US has consistently underestimated China's ability to deliver military technologies. For example, the J-20 stealth aircraft was undergoing trials and reached initial operational capability much earlier than US experts had thought likely. China is building up rapidly on issues like domination of the electromagnetic spectrum, and has strong capabilities in non-kinetic warfare. At the same time there have been difficulties with aero engines, and the radar cross-section of the J20 is probably not as small as that of US aircraft like the F22.

An overall summary of the situation has been presented in a recent issue of *Aviation Week and Space Technology* (4 June 2012). According to this report, Chinese analysts believe that if a military conflict with the US is 'confined to a short period' (say, a few weeks), China could win. This is because they could have a short-time advantage in the cyber and space dimensions. Furthermore, such an attack by China would be consistent with their doctrine of 'active defence'. According to this doctrine, if a conflict is likely China should strike first. American analysts accept that such a scenario deserves to be given credence, and agree that China has the technological capability required, but they consider other factors, such as cultural and educational, are still not strong enough in China. Weaknesses include excessive centralization, domination by the Chinese Army, inadequate expertise in handling maintenance problems, etc.

Education, Research, and Development

An important part of a good ecosystem is R&D and education. The Chinese Aeronautical Establishment (CAE), set up in the 1960s, has a staff of about 10,000 of whom 2,000 are senior researchers. Their main task is testing and development. They are spread over 20 institutes and three branch institutes, and collaborate extensively with other Chinese institutes and universities as well as with international centres and industry. Comparable aeronautical establishments in India do not match CAE in numbers. The largest, CSIR-NAL, has something like 300 to 350 qualified scientists and engineers.

Education has been recognized from the beginning as playing a crucial role in the development of the ecosystem, and has received a great deal of attention in China. There are two universities exclusively for aerospace science and technology. The Beihang University of Aeronautics and Astronautics (BUAA) has a 100-hectare campus (not unlike the Indian Institute of Science [IISc] in size), but is all devoted to aeronautics, interpreted however in a broad way. It has 17 schools, not only for technical disciplines like propulsion, aerodynamics, etc., but also for related disciplines like materials and physics, in addition to the liberal arts, law, economics, management, philosophy, foreign languages and education. BUAA has 42 institutes and 89 laboratories, a 3,300 strong faculty and staff, 26,000 students including 14,000 undergraduates, 5,000 Masters and 1,300 PhD scholars. There is no comparable institute in India; between IISc and the IITs, about a few hundred graduates and Masters are produced every year (although the number of technical colleges in the country is going up rapidly). The mantra of Beihang University is: Hard Work, Simple Living, Diligent Learning, All-round Development and Courageous Innovation—slogans that are written all over its campus.

The other university, which is at Nanjing, is almost exactly the same size as Beihang, and comparable in most respects. Both Beihang and Nanjing are administered by COSTIND. Clearly, the human resources in aerospace S&T in China are an order of magnitude bigger than in India.

Comprehensive National Power

To understand the Chinese drive towards aerospace power and the bold decisions that they keep taking, it is necessary to appreciate their national goal of acquiring comprehensive national power (CNP), a concept which they have studied at great length. Several methods have been developed to determine the sources of a nation's CNP. To illustrate the Chinese approach, we can look at the proposal of the Chinese Academy of Social Sciences (CASS). In their method, CNP is estimated as a weighted sum of indices that reflect natural resources (weightage factor 0.08), economic activities capability (0.28), foreign economic activities and capabilities (0.13), S&T (0.15), social development (0.10), military capability (0.10), government regulation and control (0.08) and foreign affairs (0.08). (The weightages add up to 1.) It will be noted that the star entries are domestic and international economic activities and capabilities and S&T, accounting between them for 56 per cent (the weights keep changing over the years; e.g. S&T has recently gone up by a few percentage points). Of course, such indices are to some extent arbitrary and reveal the thinking of the index-maker rather than providing a 'true' measure of CNP (whatever that may be). Nevertheless, the thinking behind the CASS index can be summarized in the simple Chinese formula: grow your economy, export your goods and develop your S&T, you are more than halfway there to high CNP. This philosophy has been pursued with great determination, and the results are now all too visible. Underlying this weightage is the belief that national power in foreign affairs and defence is strongly influenced

Table 4.1

Score and Rank of Different Nation-states by CNP*

Country	1989		2020	
United States	593	**(1)**	1392	(1)
USSR	387	(2)		
Germany	378	(3)	1069	(3)
Japan	368	(4)	1009	(4)
China	222	(6)	1351	(2)
France	276	(5)	669	(6)
England	214	(7)	443	(8)
Brazil	156	(8)	658	(7)
India	144	(9)	800	(5)
Canada	137	(10)	274	(9)
Australia	113	**(11)**	233	(10)

* *Source*: Predictions based on AMP method in M. Pillsbury 2000, *China Debates the Future Security Environment*. National Defence University Press, U.S.

by strengths in the economy and S&T: these will automatically lead to a rise in the country's stature and influence in the world. Indians will perhaps be surprised that social development and defence account for so little in the CNP weightage.

Using a different index devised by the Chinese Academy of Military Science, Pillsbury (2000) makes predictions about the CNP rankings of eleven countries, as reproduced in Table 4.1. It is seen that by 2020, China (1350) is expected to be very close to the US (1390) in CNP; as of now it seems to be getting there ahead of Pillsbury's schedule.

Conclusion

During the last 10–15 years, Chinese aeronautics has grown dramatically across the board—civil, military and dual technologies, all the way from stealth aircraft to unmanned air vehicles and large civil aircraft. Among these the civil sector has received greater priority recently. This shift is apparently driven by economic and technological considerations, and a keen and far-seeing appreciation of the historic shift we are now witnessing in the global aviation market from a mature West to a dynamic East. But there appears to be almost nothing that Chinese aeronautics is not pursuing: from stealth fighters to A320-type civil transport aircraft to turboprops, turbofan-engines, drones (25 new ones recently), business aircraft (assembled Cessna Citations and Embraer Legacys), and so on.

A persistent, focused, pragmatic and unwavering long-term policy; national will and ambition about China's CNP; intelligent use of the vast Chinese market as a bait that Western industry will be unable to resist; a determination to eventually design, develop and manufacture within China, learning rapidly from the presence of western industry on the one hand and the generation of a huge and well-trained workforce on the other; and the urge to challenge current market leaders: these factors have driven the growth of China's aeronautics. A remarkable feature of their recent strategy in aeronautical development has been (a) the higher priority awarded to civil aviation and (b) the pursuit of projects where they can work for and with global leaders, while simultaneously embarking on domestic projects which will challenge and eventually compete with the very same western products that they work on for now. The fact that this is feasible in civil aeronautics which, therefore, provides a convenient instrument for enhancing CNP, may well be one reason for the priority awarded to it in recent years. This poses a dilemma to western companies, but the lure of the vast Chinese market is proving irresistible. Chinese policy and actions are consistent with their theory about comprehensive national power. Thus, the C919 is intended to 'bring into full play the political superiority of the socialist system which is capable of concentrating all of its resources in achieving great things'. Since Deng Xiaoping declared some three decades ago that 'To get rich is glorious', the Chinese mantra appears to have been: 'Get rich, export and do S&T [roughly in that order], and everything else will follow.'

In the process China is quickly graduating from a low-cost 'workshop of the world' to a high-tech international competitor. They are effectively exploiting their huge, attractive and partially open domestic market, first inviting foreign investment but currently demanding technology and planning for rapid acquisition and generation of knowhow. The value of the intellectual property that China is acquiring (by this and a variety of other means) appears to be substantial.

It is therefore no wonder that Airbus CEO, Tom Enders, considers China is going to be 'the aviation nation of this century. I see no way of preventing that' (*AWST*, 12 July 2010). Airbus is now said to have one eye on Boeing and the other on Beijing in its plans for the 21st century (*AWST*, 9 July 2012). Just a few years ago no western company took China seriously in aeronautics: the rapidity with which perceptions of Chinese aeronautics have changed is dramatic. China's growing power is also seen in the way they recently threatened to stop buying European aircraft, if the European Union went ahead with a new emissions trading system (opposed incidentally by Russia and India also).

Not everything is rosy, however. The cost to the nation has included three cancelled single-engine fighters (J-9, J-12, J-13), and infructuous early attempts to make large civil aircraft (including a Boeing 707 look-alike and McDonnell-Douglas airliners). China is slowly losing its advantage in terms of labour costs, and itself faces competition from other countries. China's manufacturing technology is still not sufficiently modern. State enterprises do not appear to be sufficiently tightly managed, including in particular in the efficient use of capital investment. In addition China's demographics is weak.

The dominant role played by Asia today in civil aviation is such that a large civil aircraft that cannot or does not sell in Asia will probably not be a commercial success. Surely, there are some lessons for India in this great Chinese saga of national will to power through a complex area of high technology. The question which we have not quite asked in India is why, with the spectacular growth of civil aviation in India (in the same ball-park as China's), our national aeronautical plans are still so timid.

Note

1   COSTIND traces its origin to a civilian ministry responsible for defence procurement policy, playing also the role of a Chinese counterpart of the US DARPA. In 2008 this ministry was combined with those of Industry and IT to form the present COSTIND.

Figure 4.1

C 919

Figure 4.2

Chengdu J 20

Source: http://commons.wikimedia.org/wiki/File%3AJ20_impside_art.JPG; by Alexandr Chechin (Own work) [CC-BY-SA-3.0 (http://creativecommons.org/licenses/by-sa/3.0) or GFDL (http://www.gnu.org/copyleft/fdl.html)], via Wikimedia Commons.

Figure 4.3

Sino-Pak JF-17 Thunder

Source: http://commons.wikimedia.org/wiki/File%3APakistan_Air_Force_Pakistan_JF-17_Thunder_Ramirez-1.jpg; by Andres Ramirez [GFDL 1.2 (http://www.gnu.org/licenses/old-licenses/fdl-1.2.html) or GFDL 1.2 (http://www.gnu.org/licenses/old-licenses/fdl-1.2.html)], via Wikimedia Commons.

5

China's Nuclear Programmes

Civil, Military, and Scientific

R. Rajaraman

We are way ahead of the Chinese in developing breeders and reprocessing units needed for a closed fuel cycle strategy.

The adjective 'nuclear' is common to all three sub-topics—Civil, Military and Scientific—in the topic allotted to me in ORF's seminar on Science & Technology in China. However, in substantive terms, each sub-topic is separate and has grown into a distinct subject by itself. So, this chapter will perforce consist of three somewhat disjointed portions.

I start with the military applications, because that is the most sensitive of the three sub-topics. The term 'military' in the nuclear context means only one thing, namely, nuclear weapons. People do not like to talk about nuclear arsenals in concrete quantitative terms in broad daylight, but I will do so anyway since that is part of my mandate.

As far as official policy goes, China reiterated, in a new Defense White Paper published in 2012, its long-held nuclear policy of maintaining a minimum deterrent with a no-first-use pledge. This is roughly the same as the Indian posture, but the Chinese do not employ the qualifier 'credible' to their minimal deterrent. Beijing has never defined in quantitative terms what it means by its minimum deterrence posture, or how big a force that calls for. Of course, India too has not done that. Since there is little official information from China on its nuclear arsenal, those who try to study it have to make estimates from secondary sources. Such estimates can be inaccurate and one has always to remain willing to be corrected. In that spirit, I give in the following table estimates of the Chinese nuclear arsenal—delivery vehicles and the warheads loaded on them—taken from the Bulletin of Atomic Scientists[1] which, as far as I know, is one of the best non-governmental sources on nuclear arsenals of all countries. The table lists such data as the ranges, acronyms, and other details of the various delivery vehicles and the number of warheads carried by each. The stockpile includes land-based and submarine-based ballistic missiles as well as the aircraft capable of carrying nuclear warheads. There are about 140 land-based nuclear ballistic missiles in China, with only one warhead on each. So, that adds up to 140 warheads there. This arsenal is in a state of de-alert. This is very important and refers to the fact that the warheads are not mated with their missiles under normal circumstances and are kept separately in storage. So, I believe, are ours and also Pakistan's, unlike the US and Russia which have a large number of warheads in a launch-ready state.

There are two long-distance ballistic missiles, the DF5A and the DF 31A which are the ones that the US is most concerned about, since they are the only ones which can reach the US mainland. (India can, of course, be reached by many of the missiles in the table, if China chooses to locate and target them accordingly).

In addition to the nuclear warheads on missiles, China stores additional warheads for its submarine-launched ballistic missiles and for its Bomber aircrafts. Rumours are that there are problems with China's nuclear submarines and the missiles that sit on them. So, at present, it has probably no operational submarine launch missile capability. The basis for this comment is, again, the Bulletin of Atomic Scientists.

The total inventory of nuclear warheads in China designed for delivery by all categories of missiles and aircraft is estimated to be approximately 240. Although this number too is based on outside estimates and not taken from any official sources, interestingly this estimate has remained roughly constant for the last 10 years that I have been studying this issue. Sometimes numbers like 500 and 600 were quoted but they turned out to be wrong.

Fuller details about all the delivery vehicles and the warheads are given in the Table 5.1:

Table 5.1

List of Delivery Vehicles and Warheads Carried by Them

Type	NATO Designation	Number	Year Deployed	Range (kilometres)	Warhead × Yield (kilotons)	Number of Warheads
Land-based ballistic missiles						
DF-3A	CSS-2	~16	1971	3,1 00	1 × 3,300	~16
DF-4	CSS-3	~12	1980	5,400+	1 × 3,300	~12
DF-5A	CSS-4	~20	1981	13,000+	1 × 4,000–5,000	~20
DF-21[a]	CSS-5 Mods 1,2	~60	1991	2,150	1 × 200–300	~60
DF-31	CSS-10 Mod 1	10–20	2006[b]	7,200+	1 × 200–300?	10–20
DF-31A	CSS-10 Mod 2	10–20	2007	11,200+	1 × 200–300?	10–20
Subtotal:		*~138*				*~138*
Submarine-launched ballistic missiles[c]						
JL-1	CSS-NX-3	(12)	1986	1,000+	1 × 200–300	n.a.
JL-2	CSS-NX-4	(36)	?	~7,400	1 × 200–300?	n.a.
Aircraft[d]						
H-6	B-6	~20	1965	3,100	1 × bomb	~20
DH-10[e]		?			?	
Others?	?	?	1972–?	–	1 × bomb	~20
Total						**~178[f]**

Source: Hans M. Kristensen and Robert S. Norris, 'Chinese Nuclear Forces 2011', *The Bulletin of Atomic Scientists* 67, no. (6) (2011): 81–87.

Notes:

(a) This table counts nuclear-only versions of DF-21 (CSS-5 Mod 1) and DF-21A (CSS-5 Mod 2). The DF-21C may be dual-capable but is normally considered conventional, and the DF-21D is under development. China has a total of 75–100 DF-21s of all types.

(b) An early but limited "initial threat availability" was achieved in 2006.

(c) Neither the JL-1 nor the JL-2 SLBM is fully operational, although warheads probably are available. The JL-2 is under development but failed recent tests.

(d) China is thought to have a small stockpile of nuclear bombs with yields between 10 kilotons and 3 megatons. Figures are for only those aircraft that are estimated to have a secondary nuclear mission. Aircraft range is equivalent to combat radius.

(e) There is no clear confirmation that the DH-10 has nuclear capability, but US Air Force intelligence lists the weapon as 'conventional or nuclear'.

(f) An additional 62 warheads include those produced for SLBMs or awaiting dismantlement, for a total inventory of approximately 24 warheads.

Future Growth in Chinese Arsenals

China is the only one of the P5, the original 5 nuclear weapon states, that has not declared a moratorium on fissile material production. Although, in actual fact, it seems to have suspended production for many years, it has not come out and formally declared a moratorium, whereas the other four of the P5 have come out and done so. Clearly, China is retaining the option of increasing its nuclear arsenal sometime in the future but how quickly and by how much is not known. An US Department of Defence report also says that China is moving a large fraction of its warheads to relatively more survivable delivery systems, such as mobile solid fuel missiles which can be moved around relatively easily.

Fissile Materials

As for the production facilities and stocks of weapon-usable nuclear materials in China, some of this material (called 'Fissile Materials') has been assembled into the warheads discussed above and the rest is kept as stocks. All the quantitative information given below about China's fissile materials has been borrowed from reports of the International Panel on Fissile Materials.[2] Basically there are two species of such materials. One is highly enriched uranium (HEU), where natural uranium extracted from mines containing a mere 0.7 per cent of the fissile isotope U-235 is refined ('enriched') till it contains more than 90 per cent of U-235. This is done nowadays by using giant 'Centrifuges'. The other fissile material is Plutonium.

China has one major unsafeguarded uranium enrichment plant in Lhangzhou. That is where it has been producing its weapon-grade HEU. There are three other centrifuge units in Hangsong in the Shanshi province supplied by Russia. Their total capacity is 1.5 million SWU/year. SWU is a measure of the centrifuge's capacity to enrich uranium. To form an idea of how much enrichment capacity an SWU implies one should start with natural uranium (which consists of only 0.7 per cent of the real bomb material U-235, the rest being largely U-238), and enrich it to HEU of different higher levels of purity in U-235. Depending on whether one wants to use it for reactors, for nuclear submarines or for making weapons, it will take about 5,000 SWUs to enrich natural uranium into one bomb's worth of HEU (about 25 kilograms of HEU with 90 per cent U-235 content).

In 2010, China had begun to operate a large centrifuge plant, based on indigenous technology; all the others are Russian built.

Altogether, by now, China has a stockpile of about 16 tons, plus or minus 4 tons of weapon-grade HEU. That uncertainty of plus or minus 4 tons is admittedly large, but that is the best we know from the outside. (This does not include the roughly 4 tons of HEU spent so far in their nuclear tests and in reactor fuel submarines and so forth.) At roughly 25 kilograms a warhead, this stock of HEU can fuel 640 warheads, plus or minus 160 warheads.

To compare this with India's production of HEU, I should add that from all accounts we have not been producing significant amounts of weapon grade HEU (>90 per cent U-235). But, instead, India has been producing HEU for its naval reactors, which is believed to lie between 30–45 per cent enriched HEU. Assuming 30 per cent enrichment, our estimate is that India has produced 2 tons, plus or minus 0.8 tons, of such 30 per cent HEU. If the enrichment produced is 45 per cent, the quantity will be correspondingly less.

Plutonium is not produced in centrifuges, but is found deposited in the fuel rods of nuclear reactors and has to be extracted from them using 'reprocessing' units. China has produced 2 tons, plus or minus 0.5 tons, of weapon-grade plutonium of which about 0.2 tons must have been consumed in nuclear tests, leaving an inventory of 1.8, plus or minus 0.5 tons, of weapon-grade plutonium. A plutonium warhead roughly needs 5 kilograms, although it can be made with 4 kilograms if the bomb design and technology are really sophisticated. So, China can build an arsenal of 260 to 460 nuclear weapons just from its plutonium stockpile, in addition to what I said about uranium earlier. How much of this has been assembled into warheads, we do not know. That part is the hardest for outsiders to know.

By comparison, India's weapons-grade plutonium stocks are about 500 kilograms, plus or minus 150 kilograms. The plus or minus uncertainty comes in India's case, because we do not know the capacity factors of the reactors Dhruva and Cyrus which produced this plutonium. If the capacity and design of a reactor are known it is more or less straightforward nuclear physics to find out how much plutonium will be created; but if it is not known whether the reactor is running at 70 per cent or 40 per cent capacity, that number can vary. Hence that 'plus or minus' uncertainty in production amounts. That stock of 500 kilograms, plus or minus 150 kilograms, is worth about 70 to 130 warheads. So, some recent reports claiming that India has 80–90 warheads, whereas Pakistan has 90–100, which sent all kinds of people jumping up and down, are meaningless. You cannot estimate, from the outside, nuclear arsenal sizes with that kind of accuracy. It seems both countries are roughly in the same ball park.

To wind up this part about China's plutonium production, I reproduce a photograph I managed to get from CNTV television news. What you see in this photograph (Figure 5.1) is a cooling pond for storing spent fuel rods, and the reprocessing plant used to separate plutonium from that spent fuel.

Figure 5.1

A Cooling Pond at China's Pilot Reprocessing Plant

Source: www.news.cntv.cn (3 January 2011). Quoted in 'Global Fissile Material Report 2011', Sixth Annual Report of the International Panel on Fissile Materials, www.fissilematerials.org.

Note: This plant has the capacity of processing 50–60 tons of spent fuel per year and began opearating in 2011. The hot testing of the plant in 2010 yielded 13.8 kilograms of separated plutonium, which China declared as its civilian stockpile.

The Civilian Nuclear Programme

Most of the information given below about the Chinese nuclear energy programme has been taken by me, with gratitude, from talks given by my Chinese colleague Dr Zhu Shu Yi. He is the China National Nuclear Corporation official who comes to most international seminars on nuclear energy and is quite candid and transparent.

The first two nuclear power plants built in mainland China are at Daya Bay near Hong Kong and Quinchian which is further down south of Shanghai, whose construction started in the mid-1980s. Today, there are 13 reactors in China with a total generation of about 11 GWe (gigawatts of electric power). The distribution at present of different energy sources in China's total requirement is still mostly fossil (77 per cent) followed by 20 per cent of hydro-power and roughly 2 per cent nuclear.

Shown below is a map, *circa* 2010 (Figure 5.2), which shows the various reactor installations.

Figure 5.2

Distribution of Nuclear Power Plants in China

Source: Zhu Shu Yi, China National Nuclear Corporation, *circa* 2010.

Note: The symbols signifying 'operating', which are eleven in number, are the ones that are already working and the others are under construction, under approval etc., as shown.

As for the future, China has plans of expansion even more ambitious than India's. By 2010, they have had roughly 10 GWe , against India's 4.8 GWe, and in both countries it is roughly 2 per cent of their respective electric power generation capacity. By 2020, the Chinese expect to have 40 constructed and 20 more under construction. Percentage-wise, that will be 4 per cent to 6 per cent of the projected energy requirement for the 2020s. By 2050, they expect to have 160 GWe of electricity from nuclear sources, which will then become 10 per cent of the much higher requirement by mid-21st century, which they project to be 1600 GWe altogether from all sources. Table 5.2 summarizes all this. For comparison, the world projected number, which according to the Chinese the whole world will require or would have constructed by 2050 is 1600 nuclear GWe. These numbers have to be taken with a pinch of salt. To some extent they are still twinkles in the planners' eyes.

Table 5.2

Nuclear Power in China from 2005 to 2050

	(Nuclear Energy (GWe)	*Percent of Total Generation*
2005	8.5	2
2010	10 (India has 4.8)	2
2020	>40	4
	(60)	(6)
2050	160	10
2050 (World)	1600	

Source: Zhu Shu Yi, China National Nuclear Corporation, 2011.

By comparison, India has a nuclear generation capacity of about 4.8 GWe right now and 2 GWe more will come into stream once the two Russian-built reactors at Kudankulum begin to contribute. The following table taken from the 2006 Integrated Energy Policy report of the Planning Commission gives the planned energy mix in India. It is a somewhat old table, but a more recent one with comparable information is not available. However it does give a rough idea of our projected electrical energy estimates.

Table 5.3

India's Energy Generation Mix (GWe)

Year	Total @8%growth	Thermal	Hydro	Nuclear	Renew
2006–07	108	87.7	13.2	5.9 (actual 4.12)	1.2 Wind 0.7 (17% of 4)
2021–22	301	231	41	26.1 (little hope)	2.7
2031–32	551	430	61	57	3.6

Source: Integrated Energy Policy, Planning Commission, 2006, Table 2.7.

These projections were based on an 8 per cent growth rate, if growth rate comes down to 7 per cent these numbers will come down a bit. And as seen in Table 5.3 in 2006–07 when the report was written, against the projected 5.9 GWe of nuclear electricity, in fact we had only 4.12 GWe at that time. It has grown to 4.8 now and by 2021, which is less than 10 years from now, the total requirement is expected to go to 300 GWe, and 26 GWe of that is expected to be provided by nuclear power. Given that we have generated less than 5 GWe in the entire history of our nuclear programme since independence, this is a rather ambitious proposal. Even if we include all the planned French, Russian and American reactors and some more indigenous ones, I do not think we will reach 26 GWe by 2020. These projections were made in 2006, even before the Indo-US nuclear deal was done. So, at that time they were even more ambitious. Without the deal it would have been physically impossible to achieve this; for we just don't have the required Uranium ores. With the deal it is possible, in principle, but the rate at which things have been slowed down by the Nuclear Liability Bill and local public resistance, the 26 GWe target will probably not be reached by 2020. If I had to make a guess, I would say that we may reach, at the most, 15 GWe by then.

Turning from generation capacity to reactor technologies, most of the Chinese reactors are pressurized water reactors based largely on the Westinghouse AP 1000. They also have plans to make fast breeder reactors. On breeders, we in India are way ahead of them. The Chinese have an experimental fast reactor of about 60 MW thermal capacity while we have a 40 MW thermal fast breeder already synchronized to the grid in 1997. They plan to have a demonstration fast breeder by 2020, whereas our prototype fast breeder reactor of 500 MWe is expected to be completed soon. The earlier target of commissioning by 2010 has been missed, but I think it will probably be done by the next year. So, we are ahead of the Chinese in fast breeder technology.

The breeder reactor programmes of both China and India are part of a 'closed fuel cycle' strategy as distinct from the 'once-through' system. In the once-through system, uranium is fed into the reactor as fuel. After use, it comes out as spent fuel, and the spent fuel is disposed of in some safe storage. In the closed fuel cycle, you extract out of that spent fuel whatever plutonium has been deposited in it because of nuclear reactions in the reactor. Then the plutonium so obtained is reused as fuel, thereby extracting more energy from a given amount of the fuel you started with. In India's plan, we hope to extract U-233 from thorium placed in fast breeders and re-use that U-233 as fuel.

China takes the view, as does our DAE, that it should have a closed fuel cycle. As mentioned earlier, we are way ahead of the Chinese in developing breeders and reprocessing units needed for a closed fuel cycle strategy. So, at the 'front end' of producing the fuel China has sufficient capability to meet their requirements but at the 'back end', i.e., the end where the spent fuel that comes out is treated, reprocessed and re-fabricated into new fuel, they are behind. So, while closed fuel cycles maybe their policy they have a long way to go, much longer than India, to execute that.

At this point I would like to add a word of caution about investing too much time, money and effort on a closed cycle breeder programme. Worldwide the experience with the breeder reactors has not been an unqualified success. Around the globe there are over 400 conventional thermal reactors, but only three or four functioning commercial breeder reactors. The problem is that, unlike the normal water cooled reactors, the cooling of fast breeders is done by liquid sodium which is hot, corrosive, inflammable and eats through the pipes that carry it. Its leakage poses far more serious hazards than leakage of coolant water. The French built a breeder reactor called the Phoenix, which they felt was working very well. So they went on to build something bigger, called the Super Phoenix, but that collapsed because of repeated sodium leakage. After an expenditure of 9 billion Euros, the Super Phoenix was shut down and never ran again. Similarly in Japan, they have a breeder reactor called Monju, which again had the liquid sodium problems; they also had to close it down for many years. They started it last year but it was in trouble again later.

I do not believe either France or Japan is going to build any more fast breeders in the near future, whatever their officialdom may say to save face. Germany closed its breeder plant, whose site has since been converted into an amusement park. The US has not had a breeder for 30 years. The only country that has been using breeder reactors without complaint is Russia, but even they do not have many—just a couple. So, the number of breeder reactors that have been used successfully in the world is very small.

This is the same liquid sodium that we are going to use in our breeders too. Our DAE people believe that they have the sodium problem under control, that they have indigenously developed ways of handling it. I take their word for it. But we must think carefully about how much investment to put into our breeder programme. Previously, we necessarily had to develop the breeder-cum-closed fuel cycle programme because our uranium resources just were not enough. That compulsion is less now. Thanks to the Indo-US deal, India can now buy uranium and run conventional reactors. I know we have already invested much brain power and money into our three-stage breeder programme and I see no harm in continuing it. But it is lucky that we are not solely dependent on it anymore.

Lastly, the Chinese response to the Fukushima explosions has been qualitatively the same as ours. Neither country has abandoned its ambitious plans for nuclear expansion; they have only ordered a round of safety examinations, a comprehensive review and strengthening of safety management. However, in China all new projects, not already under construction, have been put under suspension for the moment, whereas in India, as far as I know, we are still going ahead with them.

In addition, both countries have a hand in the ITER-International Cooperation on Fusion Reactors. But fusion reactors, though slightly more real now, still remain a thing of the distant future. When I first met the great plasma physicist Marshall Rosenbluth in the 1960s, he told me that in 50 years a fusion reactor should be ready. Now 50 years later, today's experts are saying that it will be ready in another 50 years. Maybe 50 is a polite way of saying infinity! This is not to scoff in any way at the tremendous effort that has been put in by plasma physicists to set up a controlled sustainable fusion reaction. It has proved to be technologically very challenging. Predicting when it will be ready on a commercial scale is difficult. But now, especially after setting up of ITER, it looks like the present prediction of 50 years is credible. The different stages of progress from experimental to engineering to commercial scale come within range of concrete planning and timetables sound more achievable than before.

Pure Nuclear Science

In order to discuss the state of nuclear science in China, I have to enlarge the scope of the term 'nuclear science' to include sub-nuclear physics so as to bring it to today's frontiers. Investigations on the nucleus began a hundred years ago and some of the world's greatest minds worked on it. Understandably, by the 1970s already the most interesting and do-able problems dealing with the nucleus had already been studied by experimental and theoretical methods. While some good physicists, both in India and abroad, still find good problems to work on nuclear physics per se their numbers are small. Nuclear physics has not been a major cutting-edge field for decades. Most of the intellectual progeny, the descendants of nuclear physicists, have moved on to sub-nuclear physics, for which the more formal name is elementary particle or high energy physics. This technologically and conceptually more advanced outgrowth of nuclear physics probes particles *inside* the nucleus which are even smaller. The phrases that one hears in today's popular science discussions in the media and elsewhere, for example, the giant accelerator at CERN which people thought would destroy the world by creating black holes, or the Higgs or 'God' particle or String Theory—all these come from high energy elementary particle physics. This field also offers a good example which we can use to illustrate China's progress in an area of cutting-edge pure science.

One often hears that China has a very pragmatic strategy in S&T, of building a large base of technical manpower. But, as Dr Kharbanda has noted, the Chinese describe their strategy with a typical aphorism:

"Stabilize one end and free up all the rest." We have to ponder over what that mysterious phrase is meant to convey. I think what it means, as Dr Kharbanda also says, is that you do some solid basic research, and leave the rest to the market forces.

Following this philosophy, even fields like the highly esoteric subnuclear fields of particle physics and the quantum field theory, which are of no imaginable use to man or beast as of now, have been strongly encouraged. An example of this is the jewel in the crown of basic physics in China, the CAVLI Institute which is part of the Chinese Academy of Sciences. The CAVLI Institute, located at Beijing, is based on the model of the original CAVLI Institute at Santa Barbara in the USA, which is one of the top theoretical places in the world. David Gross, a Nobel Laureate and former Director of the American CAVLI Institute, is also chairing the Advisory Board of the Chinese version. In China the Institute coordinates basic research in theoretical physics in all of China and specializes in elementary particle physics, quantum field theory, string theory and cosmology. It arranges numerous workshops and conferences for all the scientific groups in China working in these fields. People from different universities and other institutions from around China are invited to spend time at CAVLI. There is continuous flow of speakers in seminars. It has only 11 permanent faculty and 12 postdoctorates, but it is still buzzing with activity. In 2010 alone, 276 papers were written by that institute, its visitors and its permanent faculty. From time to time, they have weeklong meetings where there are as many as eight talks a day. Altogether, 1,500 lectures were given in four years. That averages to about 200–300 lectures a year, an impressive number, and, by any standards, No. I.

However, in my judgement, in spite of this massive organizational effort to expose the Chinese scientists to the latest developments and experts from around the world, the quality of their indigenous output is not so high, at least in high Energy Physics and String Theory. Its impact on the world community has not so far been commensurately impressive. Whereas in India we have half-a-dozen internationally recognized hotshots in String Theory and another 20 to 30 good ones, the Chinese group is nothing like that at all. Similarly, in experimental particle accelerator physics, collaboration by indigenous mainland Chinese physicists with CERN (Geneva), the Fermi lab (the US), etc., lags behind India's.

To supplement this impression which I have gathered from the literature, seminars and conferences, a personal anecdote which illustrates the situation and, perhaps, also one of the causes behind it, will not be out of place. I had gone to Nanjing for a nuclear disarmament conference about five years back. There is a major university in Nanjing and I expressed a desire to meet the physicists there. The Vice Chancellor happened to be present at this disarmament meeting. He very generously offered to arrange a visit and it was done promptly and very graciously. I was shown around the university by a charming public relations lady, and given flowers and mementos, etc.

But I was not taken to the physics department. I gently insisted that my main interest was in going to the physics department and talking to fellow physicists there. They were not very keen about that, but out of courtesy took me to the physics building. There too, I was greeted by an official who sat me down and gave me green tea but, it turned out, he too was not a physicist but the Party representative of the department (each department of the university seemed to have one!). So, I had to persist some more and said that I wanted to meet the guys who work out equations on the blackboard with chalk, because those are my tribe. Finally, they went somewhere into the building and brought a couple of sloppily dressed but real physicists. Things improved dramatically once I met the actual scientists as compared to the layers of functionaries. I was flattered that they knew my name and had brought with them a tattered pirated copy of my book on Quantum Solutions. "Ah Professor Rajaraman!" they said and much hugging and hand-shaking followed. Finally, I was allowed to go up to their offices in the attic and happily started talking physics with them. Unfortunately, while those physicists were extremely sincere and enthusiastic, they were way behind in the choice of research problems and techniques. Nanjing University is a leading institution of scholarship in China, but somehow even there the atmosphere conducive to the flowering of original work was still being suppressed by bureaucratic structure and values.

What the Chinese do have is a large number of expatriate scientists in the western world. Some of those are top class. Further, their involvement and interaction with the mainland Chinese science groups seems to be much stronger and better organized than is the case in India. So, if the Chinese continue to persist with building an expert base and continue to give them exposure to international physics, top quality indigenous work will also emerge from China sooner or later. That should hold true for other areas of basic science as well.

Notes and References

1   Hans M. Kristensen and Robert S. Norris, 'Chinese Nuclear Forces 2011', *The Bulletin of Atomic Scientists* 67, no. 6 (2011): 81–87.
2   'Global Fissile Material Report 2011', Sixth Annual Report of the International Panel on Fissile Materials, www.fissilematerials.org.

6

IT in China*

N. Balakrishnan

Beyond contributing to economic growth, IT provides the ability to launch cyber attacks which would one day become as strategic as nuclear weapons. At some point of time cyber attack capability will become a negotiating point like the CTBT.

Introduction

The growth of IT in China is actually stunning and it forms part of China's comprehensive plan to be a world leader in technology. In India, the IT revolution started with the Y2K problem and the growth of the Web and was aided by the growing tendency across the world towards outsourcing. It was dominated by leveraging on our software development skills. Over the years, the Indian IT industry has moved up the value chain to include design of complex software systems for scientific and engineering applications and in banking. It altogether skipped the hardware component of IT and communication. But in China, IT was planned as part of a complete ecosystem of development in automobiles, space, atomic energy, aircraft infrastructure, structured education and research. The Chinese strategy in IT covered all aspects ranging from manufacturing to telecommunications to software development, software services, including consulting, managed services, education and research and also supercomputing. This is depicted in Figure 6.1.

---

* The statistics in this paper have been taken from a multitude of sources under the Fair Use Doctrine and wherever possible the sources are acknowledged.

Figure 6.1

The Complete Ecosystem for IT in China

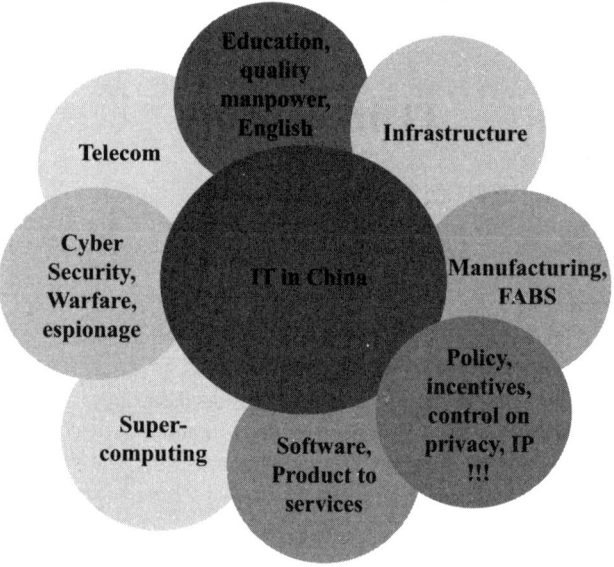

Source: Author.

There has been tremendous exploitation of Chinese diasporas' abilities, and that has enhanced the country's capability. In order to understand the IT in China, a comprehensive study of all their planning, investment and achievements is required. One would certainly conclude at the end of the study that it is a very well-planned policy that has a balance of controlled access to some of the domains to international players while China itself aggressively pursuing its reach to the world markets outside. It is also a balance between "reverse engineering" to innovative ways of capitalizing on others' 'Intellectual Property'.

In the world of software, the copying and piracy are far easier, since the software is developed in modules and could be pirated even as an executable without having to know inner intricacies of the module. The developments in Information and Communication Technology (ICT) have necessitated that almost all of the Intellectual Property of an enterprise is stored in the computers that are networked. This has resulted in intense activities of 'industrial espionage' by many countries, and China is believed and accused by the Western world to be the biggest beneficiary of this. The truth is still unclear since nothing is proved beyond doubt.[1]

China has also benefitted from the fact that it is the biggest market for ICT products and several nations clamour for this large business opportunity and hence have set up hardware manufacturing centres in China. The Chinese saga is of what a determined nation can do but may not be very easy for other nations to emulate. In the final analysis, the efforts by China to use Science and Technology (S&T) in general, and ICT in particular, as vehicles for economic growth is commendable.

China's Growth in S&T

China made large-scale investments in education and research. They built many universities that are very large even by modern-day standards. Attractive opportunities have been provided for faculty members through encouragement of entrepreneurship and through incentives for publication of papers and registration of patents. By all indicators, growth in China's science and technology is truly laudable.

The number of S&T doctoral degrees received by Chinese scholars from US universities is very large. In the year 2000, the number of US doctorates received by the Chinese was around 2,500 compared with around 800 from India. In 2009, the numbers for China increased to around 4,000 while it increased to around 1,500 for India. The number of S&T doctorates earned by Chinese students at home universities has also been going up by an order of magnitude. By 2008, China was awarding more than 28,000 doctorates in Science and Engineering from Chinese universities. China's capacity building at the high end of manpower training is also equally stunning in natural sciences and other disciplines.[2] While the IT industry in India is planning for manpower requirements in IT, it mainly concentrates on quality and numbers of the engineering graduates produced. While in China this has also included high-end manpower, including doctorates, and hence China has poised itself well for the growth at the high-end of the value chain. This has made the Chinese IT sector more robust and almost independent of the changes in policies of other countries in domains such as outsourcing.

The large investments made by the Chinese in building high-end capacity have yielded very good results. The number of papers published by Chinese scientists in the international journals and conferences, as seen in the Web of Science, is getting closer to that of the United States of America. The number of publications by scientists in China in peer reviewed international journals in 2010 was 320,800 compared with India's 71,975. China's output compares favourably with that of the United States which is around 503,000 publications. What is in favour of China is its stunning growth rate.[3]

China's publications in Computer Science alone, number around 6,000 per year of the total of around 50,000 papers per year published by the entire world (see Figure 6.2). The share of the United States in the publications in Computer Science is around 27 per cent while the shares of China and India are around 10 per cent and 1.7 per cent respectively. The trends are almost similar in every area of S&T.

While the quantity of a nation's productivity in S&T is measured by the number of publications, one of the measures of the quality is the number of times the paper is cited in other publications. The average citation per paper in Computer Science is around 6 for both India and China.[4] In view of the large number of publications by China, the total number of citations of Chinese papers is much larger than the number

Figure 6.2

Publications by Year in Computer Science

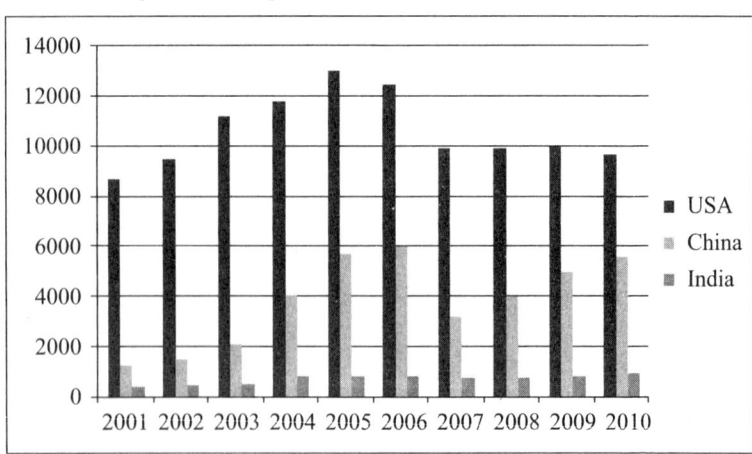

Source: Web of Science.

of citations of papers authored by Indians: thus the impact on the field by the Chinese is higher.

Another indicator of the role of S&T in wealth creation is the number of patents. While the growth rate of the number of patents created by India and China is almost similar, China continues to create almost three to four times the number of patents created by India every year, though it still is only one-third of what the US creates.[5]

China's ICT Market and Its Strategy for Investments in ICT

China's investments in ICT have been growing steadily. It doubled its investments in ICT between the years 2003 to 2008. In 2008, China's investments in ICT were at a staggering level, approximately at US$ 74 billion (461 billion Yuan) besides an investment of approximately US$ 2.79 trillion (17.29 trillion Yuan) in fixed assets.[6] As per the World Bank, the expenditure in ICT should include computer hardware, software, services and communication equipment. In such a case, the ICT investments in China in 2008 were around US$ 260 billion.[7] Today China has the largest number of mobile phone and Internet users. China is also the largest manufacturer of electronics and computer products in the world. It has also the largest market.[8]

In order to leverage on its largest ICT market, China first created a world-class scalable manufacturing base. Chinese approach to creating infrastructure like semiconductor fabrication plants, commonly called FABs, for manufacturing Integrated Circuits (ICs) is through a co-investment plan. It is also a world leader in giving very innovative incentives to induce world's biggest FABs to come to China.[9] Over the years, almost all the known manufacturers of ICs, such as TI, Intel, AMD, NVIDIA and others, migrated their bases from the west to China while the likes of Toshiba, Hitachi and NEC moved their bases from the east to China.[10] This made China the world's largest IC manufacturing base in the world. The IC manufacturing facilities were also ably supported by the assembly lines for electronic systems. It has advanced to such a state that a majority of the electronic and ICT products which one buys in the open market are manufactured in China.

In spite of the fact that there has been an excellent growth in China's semiconductor consumption market from a mere US$ 14 billion in 2000 to nearly US$ 132 billion in 2010, there is still a large gap between Chinese manufacturing capacity and its domestic needs in IC.[11] This gives ample scope for the expansion of Chinese capabilities. While exporting large quantities of electronic components and systems, China also imports to fill up the gap in manufacturing. The import is through a controlled process, and at times demands the importers to invest in Chinese facilities and increase 'Chinese IPs'. The imports and the attendant policy of investing in China have also helped a great deal in growing indigenous IP. The export is often used to leverage in foreign relationships and geographical impact.

Chinese Strategy in Capturing the World Market

When a Chinese company goes to another country, particularly in telecommunication, it faces a lot of hurdles due to the suspicion that the Chinese companies install malicious software to monitor the usage of information network and also to leak information to a Government controlled system in China. There is an abundance of such allegations in the open source.

When faced with such opposition, the Chinese do not react openly but work systematically to remove the hurdles. One such approach by the Chinese is to create joint ventures (JVs) in other countries. The other is the strategic acquisition.[12] These acquisitions have helped China amplify Chinese commercial competitive advantages and mute the Chinese commercial competitive disadvantages.

Huawei is a leading global provider of commercial telecom networks. From the company's factsheet it can be seen that "it is currently serving 45 of the world's top 50 telecom operators".[13] Huawei's products and solutions are deployed in over 140 countries and are supporting the communication needs of one-third of the world's population. As of December 2010, Huawei employed over 110,000 employees, 51,000 of whom are based outside of China. Huawei's financial results are audited on an annual basis by international accounting firm KPMG. This move is intended to gain global credibility.

In spite of all the efforts to create a sense of openness, Huawei is not welcome in the US[14] and in many other countries though it had captured many European and NATO markets merely through its cost competitiveness and extremely attractive financial incentives to the buyer. In the US, it is the suspicion that Huawei is strongly connected to the Chinese government and military that blocked its deals with AT&T, Motorola and 3COM. In the midst of such hurdles, it is commendable that Huawei has silently entered the US, and captured some markets and hired Westerners from the competition to be its spokespersons.[15] Huawei may not have broken through to the US market yet but its presence and efforts cannot be ignored for long. This is an excellent strategy to penetrate extremely hostile markets.

The Wider Fabric of Growing Chinese ICT Market and the Role of Global Players

The ICT market in the EU was estimated at US$ 1.02 trillion in 2010. The Chinese ICT market was 1.6 trillion, up 18 per cent from 2009. Growing at 18 per cent per annum, China will continue to have an ICT market which is bigger than the EU's market. "The hardware and telecommunication sector take up the lion's share of the ICT. However, software and IT services are also showing the growth potential."[16] Therefore, the Chinese are now showing very keen interest to exploit the service sector in addition to their core strength in manufacturing.

> China's telecom industry is dominated by 4 State-run telecom carriers, China mobile, China telecom, China Unicom and China DBSAT (Satellite related services). Their total telecommunication revenue in 2010 was US$ 3 trillion, up 20.5% from a year earlier. Mobile communication accounted for 70% of basic services revenue.[17]

This market is dominated by Chinese and MNC telecom equipment giants with sizeable R&D investments. MNCs have to invest in R&D in China.

The total revenue of China's software industry was US$ 203 billion, up 31 per cent from the year before. The top 100 software companies had combined revenue of US$ 37 billion in 2009 and the target is to make at least one company like Microsoft with about $80 billion revenue from software. Foreign brands still account for 65 per cent of the current software market of China because the Chinese, so far, have not concentrated much on software, but they are beginning to do it. Key domestic players in the market include Insigma, Highsoft, Hindec, Newsoft and Ufida. Entry barriers in the IT services industry in China are relatively low; especially equipment maintenance and training service are unrestricted. But for Indian companies even the IT services market is not fully open. Competition is more intense in the high-end market, and well-known foreign companies like HP, IBM, SAP, Oracle Bearing Point and Accenture have their presence in China. The Chinese have also invited Indian MNCs like Wipro, Infosys and others who have set up companies in China. Most foreign MNCs employ a large number of Chinese because of the localization needs. This, in essence, is a way by which Chinese manpower developments and training in global software development practices grow.

The outsourcing services in China are beginning to get very good rating in the international arena in spite of some concerns on intellectual property, security and privacy. It would not be surprising if one sees the emergence of a very competitive China in software development and in catering to outsourcing.

China's ICT Hubs are Geographically well spread in China

In India, the ICT centres are located mainly in a few places, such as Bangalore, Chennai, Hyderabad and Noida. In China, on the other hand, they have established ICT manufacturing centres from East to the West in many places and have aided in a balanced growth of the regional economy.[18]

China and Supercomputing

To be at the high end of ICT, it is necessary to establish world leadership in scientific computing as well, besides leadership in the electronics, PC and telecommunication market. China embarked upon a very

Figure 6.3

China's Top 10 High-Tech Cities

The main manufacturing areas for ICT hardware products are the:
• Pearl River Delta in South China,
• Yangzi River Delta in East China and
• Bohai Bay Region in North China.

Source: China Travel Guide, China Map, China Province Map, China City Map, http://www.chinatoday.com/china-map/china-map-atlas.htm

well-orchestrated plan and, today, it is amongst the top nations in supercomputing. In 2007, Indian supercomputer was the third largest in the world. Till 2007, India had better supercomputers than China. But today the fastest Chinese supercomputer is about 12 times faster than the fastest available in India, and the most alarming is the rate at which the peak performance of the Chinese supercomputers is growing year on year. In the top 10 of the 500 supercomputers in the world, there were 2 supercomputers from China in 2011 and one in 2012.[19] In the top 500 supercomputers, India had 9 in 2007 but today, it is reduced to 4, while the Chinese have 42.

IT as a Weapon

Beyond contributing to economic growth, IT provides the ability to launch cyber attacks which would, one day, become as strategic as nuclear weapons. IT can also be a weapon to conduct industrial espionage to increase a country's technological capabilities and for spying on other countries' strategic initiatives and plans. Individual hackers are also known to launch cyber attacks for their own purposes. It is believed that the cyber force in China is estimated to be around 130,000 strong. China's cyber force, strangely, was developed initially through training in the US. At some point of time cyber attack capability will become a negotiating point like the CTBT.

Conclusion

In conclusion, IT in China has been growing relentlessly. It is due to a very comprehensive and well-planned combination of investments, a comprehensive outlook to develop an entire ecosystem and novel policies. It is today robust and may not be easily shaken by any policy changes elsewhere. It is sustainable.

Notes and References

1. http://www.zdnet.com/cisco-issues-legal-challenge-to-huawei-tiptoes-us-china-dispute-7000005657/
2. Science and Engineering Indicators 2012, National Science Board, National Science Foundation, USA, http://www.nsf.gov/statistics/seind12/pdf/seind12.pdf
3. http://www.scimagojr.com/index.php
4. http://archive.sciencewatch.com/dr/cou/2009/09decALL; http://archive.sciencewatch.com/dr/cou/2009/09octCOM/
5. http://www.uspto.com; Charles Wolf Jr., Siddhartha Dalal, Julie DaVanzo, Eric V. Larson, Alisher Akhmedjonov, HarunDogo, Meilinda Huang and Silvia Montoya, 'China and India 2025: A Comparative Assessment', RAND, National Defense Research Institute, 2011.
6. Lan-Li Yi, LanZheng, Quiang Yan and Yun Li, 'The Relationship between ICT Investment and Economic Growth in China', International Conference on Advanced Management Science (ICAMS), Vol. 2 (9–11 July 2010): 136–140.
7. http://www.tradingeconomics.com/china/information-and-communication-technology-expenditure-us-dollar-wb-data.html
8. http://www.reuters.com/article/2010/08/20/china-mobiles-idUSTOE67J06020100820;http://www.internetworldstats.com/stats3.htm#asia

9. http://www.electroiq.com/articles/sst/print/volume-49/issue-3/asia-pacific/china/new-fabs-fuel-solid-growth-of-chipmaking-in-china.html
10. Marco Mora, 'China's Fast Growing IC Industry and SMIC's Role', Chief Operating Officer, Semiconductor Manufacturing International Corporation, SMIC, 2005.
11. http://www.pwc.com/gx/en/technology/assets/china-semiconductor-report-2011.pdf
12. 'Foreign Acquisitions Are Just the Beginning of China's Global Ambition', *Solid State Technology: Insights for Electronic Manufacturing*, 29 January 2013, http://www.electroiq.com/articles/sst/2013/01/foreign-acquisitions-are-just-the-beginning-of-chinas-global-amb.html; Huaichuan Rui and George S. Yip, 'The Strategic Intent of Foreign Acquisitions by Chinese Firms', *Research Highlights*, August 2012, www.IACMR.org, http://www.iacmr.org/V2/Publications/CMI/EH011101_EN.pdf
13. http://www.huawei.com/en/ucmf/groups/public/documents/webasset/hw_090305.pdf
14. http://www.reuters.com/article/2012/10/08/us-usa-china-huawei-zte-idUSBRE8960NH20121008
15. http://tech.fortune.cnn.com/2011/07/28/what-makes-china-telecom-huawei-so-scary
16. 'ICT Market in China', Report of the EU SME Centre, 2011.
17. Ibid.
18. http://www.china.org.cn/top10/2011-05/18/content_22589472.htm
19. http://www.top500.org

7

Science and Technology in the Industrial Development of China

Comparison with and Implications for India*

Ashok Parthasarathi

Due to massively wrong Government of India policies, the situation has got so bad that our indigenous technology hardware industry, which pre-1991 met 70 per cent of national needs and enabled exports on a reasonable scale now meets only 20 per cent.

Introduction

Science and Technology (S&T) have played an important role in the industrial development of both China and India right from China's liberation in 1949 and India's independence in 1947, respectively. However, the nature of that role has, at times, been similar and at other times different; so also have the respective policy orientations which have driven those roles.

The first phase of S&T in the industrial development of China was the period 1950–60. It was characterized by China near totally copying

---

* The information presented in this chapter has been gathered by the author from a large number of published sources including government publications, from conferences attended by him and in personal discussions. The interpretations and conclusions are drawn by the author.

the heavy industry model of the former Soviet Union (FSU). China's industrialization was driven by the 300 heavy industry plants in some 10 sectors set up by the FSU in China on turnkey basis. It was a phase of scientific and technological learning for China's scientists and engineers through close association with the Russian scientists and engineers, who were installing and commissioning those plants in China. All the S&T inputs in this phase were from the FSU. From several accounts in the literature, around 150 of those 300 plants were set up, most of them successfully, by the time two date-lines came up—the first was the massive dislocation of the Chinese economy caused by Mao's Great Leap (GL) forward launched in 1958, and the second was the Sino-FSU ideological split which began in 1960.

The period 1960 to 1980 made up of the Great Leap forward (1958 to 1964) followed, after a two-year gap, by the Cultural Revolution (CR) of 1966–76 saw, the total destruction of all sectors of the Chinese economy, including all areas of S&T and industry, except for the enclave of nuclear weapons and strategic missiles programmes, which were carefully and comprehensively protected by Mao.

China's Transition to an 'Open Economy'

It was only with the death of Mao in 1976 and Deng Hsiao Ping succeeding him in 1978 and quelling the mad Cultural Revolution that China returned to normalcy around 1980. However, what Deng had bequeathed was a totally shattered economy. After much discussion at many levels of the Chinese Communist Party, it was decided to make a total break with the policy of an inward looking economy, and to 'Open up to the World' in industry and foreign trade. Foreign Direct Investment (FDI) across the board, including in Consumer Goods, and many areas of internal trade were massively liberalized. With these huge changes, large amounts of foreign technology, both as a part of FDI and independently from it, flowed into China. Western multinational companies also went into China in a big way. Combined with a high savings rate of 40 per cent of GDP and a large volume of labour-intensive industry, both the domestic market and exports grew phenomenally. GDP grew at 9–10 per cent on a sustained basis for 20 years from 1985 to 2005 By 2005, the share of industry in GDP had reached the phenomenally high figure of around 50 per cent.

The Indian Transition

India launched its industrialization programme as the core of its second Five-Year Plan (1955–61). However, there were two distinctions with respect to the Chinese case. First, the fact that from 1956 onwards, India had economic and technological relations not only with the market economies of the West, but also with the FSU and other centrally planned economies of Eastern Europe. Second, India chose to adopt a mixed economy with both public and private sectors in industry and trade, both internal and external. India operationalized these distinctions with public sector industry directed and financed by the State, and the private sector from 1955 to the opening up in 1991 regulated by the State and financed by a mix of the State (public sector banks and financial institutions) and the market.

Somewhat similar to China, India built a good part, though not all of its core industries with technology and plant and equipment provided by the FSU and the communist countries of Eastern Europe. However, with the maturation of its own technology and machine and equipment design, engineering and manufacturing base, India built more and more of its industrial plants and products on its own. But despite such maturation, India continued to depend on technology imports for both its public sector and private sector industries and plants. From 1960 to 1990, India's savings rate was, like China's in the latter's pre-opening-up phase, in the range of only 22–26 per cent. Its GDP growth rate in the 1960s was in the 5–6 range, but dropped to 3–4 per cent in the 1970s. With partial deregulation, the growth in the 1980s rose to 5–7 per cent.

India faced up to the 'Oil Shocks' of 1973 and 1979 remarkably well with its Balance of Payments remaining positive throughout the period. However, this was not so in the case of the oil shock of 1991, when foreign exchange reserves plummeted to such low levels that they were only adequate to finance three months of imports. To combat this problem and to put the country on a substantially higher growth path,

India made a radical break with its earlier economic policy by 'opening up' in 1991, much as China had done about a decade earlier. The policy elements of that 'opening up' of India were also similar to those of China. After a few years of economic dislocation at the turn of the 1990s India also launched itself on a substantially higher growth path on a sustained basis until 2010.

It is against this background that S&T in China's industrial development and the comparison of India's strategy and experience with that of China are presented in the rest of this chapter. The framework adopted is one of presenting and analyzing the following set of key sectors of Industry in the two countries.

- Steel
- Fossil fuel-based Electric Power
- Electrical Power Plant Equipment
- Renewable Energy
  - Wind Power
  - Solar Energy
- Telecommunications
- Pharmaceuticals

Steel Industry

Historically, the foundations of China's steel industry were laid by imperial Japan when it colonized Manchuria, in the 1930s.

There was then almost a two-decade hiatus until the People's Republic of China refurbished those old Japanese steel plants and also set up its first modern steel plants in Manchuria and the central Chinese city of Wuhan using Soviet technology and plant and equipment on a turnkey basis in the late 1950s.

In 1958, giant hoardings all over China quoting Mao's goal that Chinese steel output would exceed the then UK output of 24 million metric tons (mmt) per year by 1965, were very common. However, Mao's disastrous Great Leap programme of 1958–61 in which millions of backyard pig iron-making crude furnaces established by 'the people' were the central symbol, set back a modern steel industry in China.

Figure 7.1

Growth of China's Steel Output

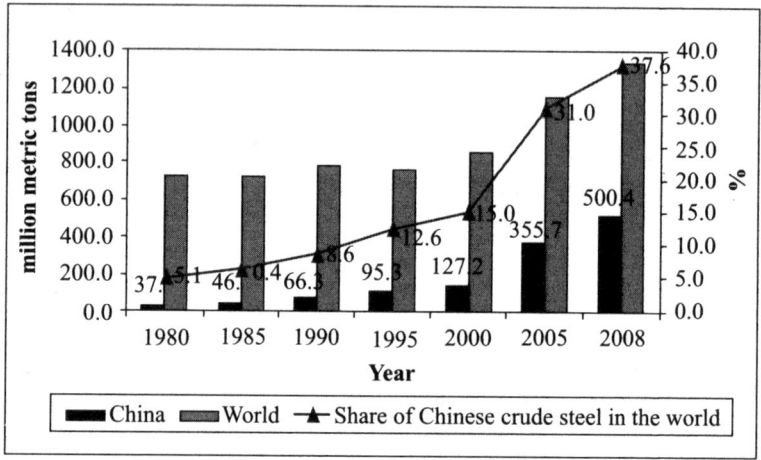

Source: The World Steel Association, 2009.

The following of the Great Leap by the Cultural Revolution, starting in 1966 and mercifully ending with Mao's death in 1976, practically totally destroyed the whole modern industrial sector in China, including a modern steel sector. That goal was painstakingly achieved all over again from 1980 onwards.

The growth of steel output over 1980–2008 against the backdrop of World Steel Output is shown in Figure 7.1.

It is seen from the figure that the Chinese steel output grew from 37 million metric tons in 1980 (5 per cent of world production) to 500 million metric tons in 2008 (38 per cent of world production). The astonishingly steep increase from 120 million metric tons in 2000 to 500 million metric tons in 2008 is particularly notable.

Table 7.1 shows the world's 'top 10' steel-producing countries in 2008. China's no. 1 position at 500 million metric tons vis-à-vis only 118 million metric tons of Japan and 91 of the US is the first indicator that there is something seriously wrong with China's steel industry.

And indeed there are two major illnesses; namely, over production and industrial fragmentation.

Table 7.1

World's Top 10 Steel-producing Countries, 2008

Country	Rank	Production 2008 (million metric tons)	World Share (%)
China	1	**500**	37.6
Japan	2	118	8.9
US	3	91	7.2
Russia	4	68	5.2
India	5	**55**	4.1
South Korea	6	53	4.0
Germany	7	45	3.4
Ukraine	8	37	2.8
Brazil	9	33	2.5
Italy	10	30	2.3

Source: The World Steel Association, 2009.

Over Production

As for the first malaise of this production, a detailed input-output analysis by the Economist Intelligence Unit (EIU) of London which came out in mid-2010 indicates that, despite having a widespread and diversified industrial base, almost all sectors of which were significant steel consumers, total domestic consumption in 2009 was a mere 370 million metric tons, whereas steel production was a highly excessive 600 million metric tons.

The EIU study concludes:

> Despite central planning and constant oversight by the Chinese National Reconstruction and Development Commission, the steel enterprises of China are very highly wasteful, using too much iron ore, coke, electricity, manganese and other reductants per tonne of finished steel and also overall in the aggregate making too much steel.

How did that happen? By China's national drive in all industrial sectors to be a quantitative numero uno at the world level—quality, efficiency and environmental impact be damned!! Also, without any reference whatsoever to real/effective demand.

Industrial Fragmentation

The second terrible illness of China's steel industry is that it is the most fragmented in the world. It has around 7,000 iron and steel companies— far more than any other country. This may be partially seen from Table 7.2, which also shows the productivity of China's steel-making enterprises vis-à-vis that of the world's leading firms as of 2007.
Moreover, even Table 7.2 accounts for only 76 of the total 7,000 steel plants in China. Secondly, the complementary data from Chinese sources reveals that the 'bottom' 1,000 iron and steel 'plants' are producing as little as 100 tons/year!! Thirdly, none of those 100 TPA 'toy' plants of

Table 7.2

Productivity of China's Large Steel-making Enterprises and the World's Leading Firms, 2007

Enterprise	Production of Crude Steel (mmt)	Total Employees (In Thousands)	Productivity (Output per employee) (tons)
China			
Baosteel Group	28.5	40	712
Angang-Bengang Group	23.5	196	119
Jiangsu Shagang Group	22.8	26	876
Tangshan Iron & Steel Group	22.7	96	236
Wuhan Iron & Steel Group	20.1	87	231
Shougang Corporation	15.4	80	192
Magang Group	14.1	59	238
Jinan Iron & Steel Group	12.1	41	295
Laiwu Iron & Steel Group	11.6	39	297
Hunan Hualing Iron & Steel Group	11.1	46	241
Other 66 Iron & Steel Groups	Less than 10	n/a	n/a
Comparison with the World			
Nippon Steel (Japan)	35.7	14	2550
JFE (Japan)	34.0	14	2428
POSCO (South Korea)	31.1	17	1829
U.S. Steel (U.S)	21.5	21	1023

Source: The World Steel Association, 2009.

Table 7.3

Coke Consumption per Ton of Crude Steel Output of China, India and Other Leading Steel-producing Countries (in kilograms)

Country	2005–06	2006–07	2007–08
China	740	720	710
India	705	645	586
Japan	380	380	420
South Korea	230	190	180
The U.S.	200	170	170

Source: The World Steel Association, 2010.

gigantic inefficiency have been closed down by the Chinese government, because they are municipal-level plants each employing huge numbers of workers.

As far as techno-economic efficiency is concerned,, Table 7.3 shows that in a ranking of the steel industry in the five major steel-producing countries, including India, the Chinese industry is at the very bottom of the heap in terms of a key parameter viz., coke consumption per ton of crude steel. While in 2005–06, the average consumption of coke in Chinese steel plants was 740 kilograms per ton of crude steel produced, and the Indian average was only a wee bit less than that at 705 kilograms per ton, by 2007–08 the corresponding figures were 710 and 586 kilograms per ton i.e., the coke consumption in the Indian steel industry was a substantial 125 kilograms or 15 per cent lower than in the Chinese steel industry.

In regard to energy consumption in their respective biggest and most efficient plants, it was 20 per cent higher in Chinese plants compared to Indian equivalents. It is the same story in regard to by-product recovery.

The Indian Steel Industry: Structure, Growth and Technology

The steel industry in India was launched by the pioneering 300TPA plant set up by J.N. Tata at Jamshedpur in today's Jharkhand in 1935 with technology supplied by the British Steel Works in Sheffield in the UK.

Table 7.4

Energy Consumption Data of Indian Steel Plants

Sl No	Plant	1998–99	2008–09	% Reduction
1	Bhilai Steel Plant (BSP)	7.16	6.50	9.22
2	Durgapur Steel Plant (DSP)	7.87	6.51	17.28
3	Rourkela Steel Plant (RSP)	10.49	7.09	32.41
4	Bokaro Steel Plant (BSL)	8.50	6.83	19.65
5	IISCO Steel Plant (ISP)	10.85	8.18	24.61
6	Steel Authority of India Ltd. (SAIL)	8.17	6.74	17.50

Source: Performance of Indian Steel Computed by the Author from Industry, and Ministry of Steel, Government of India, 2010 and World Steel Association, 2010.

Note: Best Chinese Plant: Bao Steel Group % reduction 1998–99 to 2008–09: 9.67.

Then, as the core of the second Five-Year Plan 1956–61, the Government of India set up three 1 million ton per annum steel plants at Bhilai, Durgapur and Rourkela on a practically turnkey basis by the FSU, the UK and the then West Germany, respectively. While the first two were based on the traditional Open Hearth Technology, the Rourkela plant used the then state-of-the-art LD converter process. The Bhilai and Rourkela plants were completed on time and to cost, even if both, like Durgapur, were financed by credits by the technology supplying countries. However, the UK-assisted Durgapur plant gave major technological problems from the word go. The cost per ton was in the following order: Bhilai (FSU) (lowest), Rourkela (WG) next and Durgapur a long way up. So was the case of electricity and coke consumption per TPA and quality per TPA. This demonstrated that despite the Rourkela plant being based on the 'latest' technology, the FSU was able to match, and, in some performance parameters, exceed the West German technology.

As the years went by, both Bhilai and Rourkela were steadily expanded to their present capacities of 5.7 and 2.3 MTPA, while Durgapur became the sick child of our steel industry. Equally, if not more importantly, both the production process and the plant and equipment in all these plants became totally indigenized and improved in terms of production process efficiency, product range, diversification and cost.

In 1962, a fourth steel plant, this time with a starting capacity of 4 MTPA came up at Bokaro with technical assistance (not technology) from the FSU and with 80 per cent indigenous plant and equipment. Today, this excellent plant has a 9 MTPA capacity with hugely low coke and electricity consumption matching world levels.

In 1968, the FSU set up a Central Engineering and Design Bureau (CEDB) for the steel industry to enable us to design, engineer, diversify and expand our steel industry on a totally self reliant basis. In 1974 CEDB was converted into a national consultancy and design engineering company called Metallurgical Consultants Ltd (MECON). That year also saw the integration of all the four public sector plants into a single company called Hindustan Steel Ltd (HSL) and a few years later into separate companies under a holding company called the Steel Authority of India (SAIL).

In very sharp contrast to the Chinese iron and steel industry dealt with earlier, the Indian iron and steel industry consists, as of 2010–11, of a relatively small number of 11 steel plants of medium size as may be seen from Table 7.5.

Table 7.5

Indian Iron and Steel Industry (as of 2010–11)

S.No.	Plant	Size Million Tonnes (MT)
1	BSP, Bhilai	5.70
2	BSL, Bokaro	4.10
3	DSP, Durgapur	2.14
4	RSP, Rourkela	2.30
5	ISP Burnpur	0.49
6	Alloy Steel Plant (ASP), Durgapur	2.40
7	Stainless Steel Plant (SSP), Salem	3.00
8	Rashtriya Ispat Nigam Ltd (RINL): Vizag	3.83
9	**TATA Steel Jamshedpur & Corus, Belgium**	7.52
10	JSW	4.76
11	JSPL	1.20
	TOTAL	**55 MTPA**

Source: Ministry of Steel, Government of India, 2011.

Note: Eight of the above are in the public sector (seven with SAIL) with a current total capacity of 15 MTPA while the remaining three are in the private sector.

Table 7.6 shows the comparative productivity of China's and India's large steel-making enterprises and the world's leading firms, in 2007.

Table 7.6

Productivity of China's, India's Large Steel-making Enterprises and the World's Leading Firms, 2007

Enterprise	Production of Crude Steel (mmt)	Total Employees (1,000)	Output per Employee (tons)
China			
Baosteel Group	28.5	40	712
Angang-Bengang Group	23.5	196	119
Jiangsu Shagang Group	22.8	26	876
Tangshan Iron & Steel Group	22.7	96	236
Wuhan Iron & Steel Group	20.1	87	231
Shougang Corporation	15.4	80	192
Magang Group	14.1	59	238
Jinan Iron & Steel Group	12.1	41	295
Laiwu Iron & Steel Group	11.6	39	297
Hunan Hualing Iron & Steel Group	11.1	46	241
Other 66 Iron & Steel Groups	Less than 10	NA	NA
Comparison with the World			
Nippon Steel (Japan)	35.7	14	2550
JFE (Japan)	34.0	14	2428
Posco (South Korea)	31.1	17	1829
U.S.Steel (U.S)	21.5	21	1023
India			
Bhilai Steel Plant	5.055	34407	1470
Durgapur Steel Plant	1.914	14743	1330
Rourkela Steel Plant	2.093	21105	1000
Bokaro Steel Plant	4.127	28978	1333
Alloy Steel Plant, Durgapur	0.157	NA	NA
Indian Iron & Steel Plant	0.458	NA	NA

(continued)

Table 7.6

(continued)

Enterprise	Production of Crude Steel (mmt)	Total Employees (1,000)	Output per Employee (tons)
Viswaswaravya Iron and Steel Ltd	0.159	NA	NA
Tata Iron and Steel Company	5.01	NA	NA
Rashtriya Ispat Nigam Ltd	3.32	NA	NA
Special Steel Plant Salem	0.565	NA	NA

Source: The World Steel Association, 2011.

Table 7.7 shows coke rate per ton of crude steel of Indian steel plants in comparison with the best and largest Chinese plant.

Table 7.7

Coke Rate per Ton of Crude Steel of Indian Steel Plants

Financial Year	DSP (T/T)	BSP (T/T)	RSP (T/T)	BSL (T/T)
1997–98	0.72478	0.60141	0.82222	0.67206
2007–08	0.59566	0.53045	0.60294	0.61626

Source: Steel Authority of India Ltd (SAIL), 2009.

Note: Best Chinese Plant: Bao Steel Group, 0.86 T/T in 2007–08.

Electrical Power

The Electric Power Utility Industry

Around 70 per cent of China's electric power utility industry is coal-based, 20 per cent is hydro, 5 per cent is nuclear and 5 per cent is renewable. The dominant feature of the industry has been one of huge shortages: brownouts and blackouts from as far back as the early 1990s. The average electricity availability shortfall in the 1990s was 10–15,000 MW. However, in 2004, the shortfall rose to around 20,000 MW.

By the summer of 2011, the shortfall had increased further to 35,000 MW. The problem's magnitude is well reflected in the fact that over 2005–10 China's total actual installed power generation capacity increased by 445,000 MW i.e., on an average of around 90,000 MW per year. Despite these huge increases—more, annually, than the total increase of 65,000 MW during India's Eleventh Plan as a whole and only a little below India's targeted production of 100,000 MW during the Twelfth Plan (2012–13 to 2017–18)—blackouts in China have not been eliminated.

Senior Chinese government policy makers and managers say 'many factors' have been responsible for such a situation:

First, cut-back on investment on expansion of existing, and setting up of new power plants from around 2005: the ageing of Chinese power plants leading to steep reductions in Plant Load Factors (PLFs) and Power Efficiency and the consequent low total power output from these ageing and inefficient plants: frequent plant breakdowns, generation interruptions at the power station/plant level and grid network collapses.

Second, high dependence on coal-fired power plants was seriously affected by shortages in coal production; thus leading to a steep increase in coal prices. Over just 2007 and 2008, coal prices shot up 300 per cent, which consequently caused a 15 per cent increase in power costs. But the Chinese Government took a policy decision not to increase power prices, nor to shut down the oldest thermo-electrically most inefficient and environmentally most 'dirty' and very high carbon footprint plants. This, in turn, meant that the utilities did not have the cash needed to buy and install additional new power plants. It was this 'chain' which directly triggered the severe electricity shortage of 35,000 MW nationwide in 2011.

Third, according to the Chinese National Statistical Bureau, in Zhejiang Province, in China's heavy industrial concentration with Shenyang city as its capital in the north-east, the industrial value added was 14 per cent year-on-year, whereas the output of the energy intensive sub sectors, for example, metallurgy and heavy chemical industries located in the same region, soared by 30 per cent.

Fourth, regional imbalances: the 'high economic gross output' eastern Chinese coastal areas lacked both coal and power plants, whereas the economically under-developed western and northern regions had both, but no industrial loads. Moreover, the weak inter-regional power transmission grid led to inter-regional power transfer being severely limited by transmission capacity problems. In this respect, India is far better off due to the setting up of a large amount of power generating capacity in the west, north and south in addition to large pit-head coal-based plants in the coal heartland of the east. Moreover, the considerable hydro-capacity in the north and northeast is being aggressively exploited. Equally important, the Government of India set up, over 12 years ago, the Power Grid Corporation of India Ltd (PGCIL), whose excellent performance has enabled India to tackle the inter-regional power transfer problem very well. Concurrently, the setting up of state, regional and the national Load Dispatch Centres, all wholly on the basis of indigenously developed technologies and consisting of both land and satellite based voice and data communication channels (at less than 50 per cent the cost of equivalent systems if imported from companies in western Europe or Japan), and linking the state-of-the-art (again wholly indigenously developed) computerized power management centres on a real time/on-line basis have saved India from China's serious problems in this area. PGCIL's national high voltage AC and DC power transmission systems are also wholly indigenously designed, engineered and set up using indigenously made advanced technology equipment. It is currently operating as a 400 KV system but is in the process of being upgraded to 765 KV, a task to be completed by end-2013. This puts India ahead of China in this critically important field.

The Electric Power Generating Equipment Industry

About three years ago, the four top Chinese electrical power generating equipment manufacturing companies led by China's largest company in this field viz, Dong Feng Power Plant Manufacturing Company, tried to get into our electric utility industry. The prices at which they offered to supply power generating equipment to our state electricity boards/power departments were at least 30 per cent lower then those of our power plant manufacturing companies led by our jewel, BHEL. This was largely because the cost of the power plant equipment the Chinese companies were proposing to supply was not only 30 per cent but in some cases as much as 40 per cent lower than world prices i.e., they were dumping!

This meant that the 10 per cent customs duty on thermal power plant equipment prevailing at the time in India was grossly inadequate for our national power plant manufacturer, BHEL (production capacity 20,000 MW of power plant equipment per year compared to Dong Feng's only 10,000 MW per year) to compete with the Chinese on the basis of a level playing field.

So, BHEL and its parent Department of Heavy Industry (DHI) sought from the Finance Ministry an increase in the customs duty from 10 per cent to 30 per cent or the imposition by the Commerce Ministry of an Anti-Dumping Import Duty on the Chinese companies of 50 per cent. However, the Power Ministry of the Government of India lobbied hard for the customs duty to remain at the prevailing 10 per cent level on the mistaken notion that our power generating utilities would then be able to get Chinese equipment at much lower prices, and so the utilities would be able to generate and sell power at much lower prices than hitherto from BHEL without any sacrifices in performance. Finally, on the intervention of the Planning Commission, a compromise about increase in customs duty from 10 per cent to 20 per cent was agreed upon. While this inter-ministerial battle was going on, the Chinese companies managed to get their first few orders from some state electricity boards (SEBs), mostly for power plant equipment of smaller 250 MW size. They also went ahead with site preparation, equipment import, installation and commissioning over the next 24–30 months.

By mid-2008, the very first Chinese power plants got connected to the national grid. Their initial operations were passable, except for several equipment interruptions and even failures. But by early 2011, serious equipment failures began to occur. So, several SEBs and other state power utilities cancelled their planned further orders on the Chinese companies and returned to the technologically and operationally well-proven BHEL.

A particularly serious operational and equipment problem, which the Chinese did not take advance action to combat, was the fact that our coals have the highest ash content of all thermal coals worldwide. So, when they are burnt in the boilers of electrical power plants, they rapidly and significantly erode the walls of the boilers making the boilers fall apart. Long used to this, BHEL had used designs, engineering and special materials of construction for the boilers made by it which made them totally erosion proof for decades. The Chinese, in contrast, have not been able to develop the same boiler technology as BHEL and so they had to replace the boilers in their coal-fired power plants frequently adding a substantial unplanned capital cost to their power plants, which also meant repeated and prolonged power station shutdowns as the boilers were changed.

The technological weaknesses and deficiencies of Chinese-made turbines and generators (TGs) are also great in comparison with the sustained high performance, quality and reliability of BHEL's TGs. For example, secondary fuel oil (SFO) consumption of Chinese turbines and generators are, according to an independent report by JM Financial and Merril Lynch Consultants, 12 times more than in BHEL turbines. Consequently and after independently verifying the Merrill Lynch result, our Comptroller and Auditor General (C&AG) in his recent report on the Electrical Power Equipment Industry has pointed out that several of our SEBs and state generating companies (GENCOs), which had gone in for Chinese turbo generators, had to suffer serious financial losses due to the hugely higher SFO consumption thereby leading to avoidable massive loss of public funds. The C&AG report further states that, despite the Chinese power plants being initially 15 per cent cheaper in terms of capital costs, the life cycle cost of BHEL equipment was 30 per cent lower due to the BHEL equipment having much lower operating costs, better plant load factors (PLFs), plant efficiencies (PEs) and extremely low down times and operating interruptions, much longer time between overhauls, etc. The Merrill Lynch report states: "BHEL retains a 15–20% edge over the Chinese thermal power plant manufactures due to lower PLF, heat rate and auxiliary consumption." Moreover, the same report notes that all the Control Electronics and Instrumentation in BHEL power plants are two technological generations ahead of those in the latest Chinese power plants offered by Dong Feng.

Another, key technological issue is the 'criticality level' of the thermal power plants of the two countries. Up till 2005, all our and Chinese power plants were based on sub-critical technology in terms of the coal burnt in their power plant boilers. However, in 2006, a policy decision was taken by the Government of India that power plants to be commissioned from 2011–12 should be based on Super Critical Technology for plant capacities of 660 MW and higher power output levels up to 1000 MW units, because Super Critical Technology leads to 15 per cent better efficiency and cost of power generated. The first 660 MW super critical power plant set up in India was a plant of the largest power utility company, the National Thermal Power Corporation (NTPC) located in Talcher in Orissa and wholly imported from Russia on the basis of International Competitive Bidding (ICB). It was commissioned in November 2010. At the same time, a BHEL-developed super critical power plant on the basis of joint design, development and engineering by a consortium of BHEL with GE, Alsthom and Siemens was also launched. The first 660 MW super critical plant from the consortium was successfully developed and has been undergoing extensive and rigorous tests and evaluation at the Vijayawada thermal power station. It was expected to be connected to the grid by end-2012. Chinese thermal power plants, even the ones most recently operationalized, are still based on the older and less efficient sub-critical technology; and, what is more, no plans have been announced by the Chinese Government to-date to upgrade their power plant equipment industry to super critical technology.

The third was one of a major 'discontinuous'/fundamental technological advance and not an incremental one; viz, a near-operational scale of Integrated Gasification Combined Cycle (IGCC) Power Plant. In IGCC technology, the usual fuel of pulverized coal as boiler fuel is wholly replaced by clean coal gas from a gasifier. This technology enables securing: (a) 5 to 10 per cent higher efficiency than even the Super Critical Technology-based power plants and a likewise consequent reduction in the cost of power produced of 10 to 12 per cent; and (b) it also totally eliminates all air polluting emissions from the plants, thereby steeply reducing the carbon footprint of coal-based power plants apart, of course, from eliminating the adverse effects on human health.

China

A joint venture company between four Chinese coal-based power equipment manufacturing companies and Peabody Energy Inc of the USA is in the process of setting up China's first IGCC power plant of 650 MW capacity at a cost of US$ 5.7 billion based on Peabody's technology. The cost of this project works out to ₹ 30, 000 crores at an exchange rate of one US$ = ₹ 50 which comes, in turn, to ₹ 46 crores per MW! The start-up year, according to estimates in the latest literature, is projected as 2014–15. More importantly, it is being set up as a totally turnkey plant. From a time-of-completion viewpoint, the Chinese have absolutely no control over it. This is established by the fact that the project is three years behind the originally committed schedule of Peabody viz., 2011–12. Now, 2014–15 is the completion year as committed by Peabody. What is more, top Chinese project managers say even 2014–15 is, at least, a one-year, and more likely, a two-year under-estimate with consequent huge cost escalations.

India

In contrast to the above Chinese status in regard to its first IGCC Plant, the Indian IGCC programme is far superior in many respects. First and foremost, the Indian programme is based on wholly indigenous Indian technology developed by BHEL. Second, the Fluidized Bed Gasifier Technology chosen for the plant is one in which BHEL (the sole turnkey executing agency) has three decades of knowhow and experience. This experience was built up because of the need to use India's (unlike China's) high ash coals (China's coals have high sulphur). The US and Europe use Entrained Bed Gasifier Technology because their coals have no ash. Third, after detailed study of the IGCC technology flowsheet, BHEL first designed and developed a 6 MW pilot IGCC plant, ran it for two years and optimized the technology on the pilot plant not just technically but techno-economically as well. Based thereon, the technology is being upscaled to a 200 MW commercial plant with BHEL generating the complete basic process, design and engineering packages. Concurrently, meticulous costing was done and a figure of ₹ 1,600 crores, or only ₹ 8 crores per MW was arrived at. A financial partnership was also struck by BHEL with the Andhra Pradesh State Electricity Generating Company (APGENCO), and a site survey led to the choice of the Vijayawada Plant premises of APGENCO which the latter is providing free of cost. An overall financing package was worked out in the 3rd quarter of 2011 involving NTPC, APGENCO, BHEL, the Department of Heavy Industry and the Ministry of Power to fully cover the ₹ 1,600 crore project cost.

Approximately, a 100-engineer project team of BHEL-NTPC-APGENCO was assembled and ground-breaking took place on 2 January 2012 with the project completion date set on 30 June 2015 i.e., a three-and-a-half-year project duration. The project team has a high confidence level that the project will be completed within the approved cost and time targets.

The massive contrast between the Chinese IGCC plant's total project cost per MW of ₹ 46 crores and the Indian plant's cost of ₹ 8 crores per MW is unprecedented. It is very largely due to the fact that the entire knowhow and capital equipment involved in the Indian plant is indigenous and that a US transnational has taken the Chinese for a ride. The overall conclusion, therefore, is that in the area of IGCC coal-based electrical power plants, India is far ahead of China in terms of: (*a*) costs—both CAPEX and operating; (*b*) technological knowhow; (*c*) indigenous vis-à-vis foreign technology; (*d*) near 100 per cent indigenous plant equipment; (*e*) overall plant generating efficiency—44 per cent compared to 40 per cent; and (*f*) effectively zero air pollution/emissions.

Renewable Energy Overview

China launched its overall programme in various renewable energy sources—wind, solar, biomass, etc.—around 2005 with the promulgation of a comprehensive Renewable Energy Law in that very year. However, policies for and programme in individual renewable energy sources were launched much earlier, for example, wind energy as far back as 1988 and solar energy in 1998. On 18 December 2009, the Chinese government made a political commitment to the international community at the Copenhagen Conference on Climate Change that non-fossil energy would satisfy 15 per cent of the country's energy demand by 2020. This goal became a binding target for short-term and medium-term national social and economic planning, together with a subsequently formulated target that CO_2 emissions per unit of GDP would be 40–45 per cent lower in 2020 than in 2005. On 26 December 2009, the Renewable Energy Law was amended. The new law required grid operators to absorb the full amount of renewable power produced, also giving them the option of applying for subsidies from a new 'Renewable Energy Fund' to cover the extra cost related to integrating renewable power, if necessary. Grid operators refusing to buy power produced by renewable energy generators could be fined up to double the loss suffered by the renewable energy generator.

Timelines of key policies and measures for the development of the renewable energy sector between 2005 and 2011 are set out below:

2005
- Renewable Energy Law
- Renewable Energy Industry Development Guidance Catalogue

2006
- Trial Measures for Management of Cost-Sharing in Renewable Energy Power Generation

2007
- Management Rules related to Renewable Energy Power Generation
- Mid- to Long-Term Development Plan for Renewable Energy

2008
- Renewable Energy in the eleventh Five-Year Plan (2005–11)
- Finance Ministry Notification on Interim Measures for Management of Special Project Funds for the Industrialization of Renewable Power Generation Equipment

2009
- 15 per cent non-fossil fuel energy target set for 2022
- 40–45 per cent Carbon Intensity Reduction Target set for 2022
- Perfect Policies on Grid-Connected power Pricing

2011
- The Emerging Overall Energy Industry Plan
- Twelfth Five-Year Plan of Renewable Energy Development (2012–17)

An assessment of the progress of the renewable energy sector as a whole in China would not be possible in this chapter because of the paucity of available data. Suffice it to say that the progress on most counts has been quite good; indeed in the two sub-sectors dealt with in detail in the following pages viz., Wind and Solar Photo-Voltaic (SPV), the progress has been spectacular.

To turn now to India's R&D, work on renewables, especially SPV, had started as early as 1977 at the National Physical Laboratory and IIT, Delhi with the industrial work launched at the public sector company—the Central Electronics Ltd (CEL), situated near Delhi. In 1980, the Planning Commission provided funding to the tune of ₹ 50 million (US$ 5 million) to CEL to set up a Demonstration Plant to manufacture 1 MW per year of solar cells and panels as also design, engineer and volume-produce a range of SPV systems using those panels for extensive user trails in the field. This was called NASPED i.e., National Solar Photo-Voltaic Demonstration Programme.

In 1982, the Government of India set up the world's very first government agency to deal with all aspects of all non-conventional energy sources. It was called the Department of Non-Conventional Energy Sources or DNES. It was under DNES that all renewables came to receive intense and integrated support of all kinds—promotional and regulatory—from human resource development through design and development of technologies and products/systems, resources assessment studies for wind, solar and biomass, demonstration projects, funding of commercial projects through its public sector financing agency, the Indian Renewable Energy Development Agency (IREDA) and a nodal R&D product-testing and evaluation and certification centre called the Solar Energy Centre, situated near Delhi.

Figure 7.2 sets out the composition of the Indian electrical power sector as of March, 2012. While Table 7.8 presents the renewable energy-based electrical power potential in India as of the same year. Table 7.9 indicates how renewables have grown in the most recent tenth and eleventh Five-Year Plans. Particularly important is the fact that the Government of India launched, in January 2010, a 12-year Jawaharlal Nehru National Solar Mission with the goal of producing 22,000 MW of solar energy-based electric power generation by end-March 2022, out of which, 20,000 would be grid connected while the remaining 2,000 MW would be stand-alone. If this goal is achieved along with comparably ambitious goals in wind and, to a lesser extent, bio power, the share of renewable power vis-à-vis the entire range of conventional power sources, which was

Figure 7.2

Indian Electrical Power Sector at a Glance

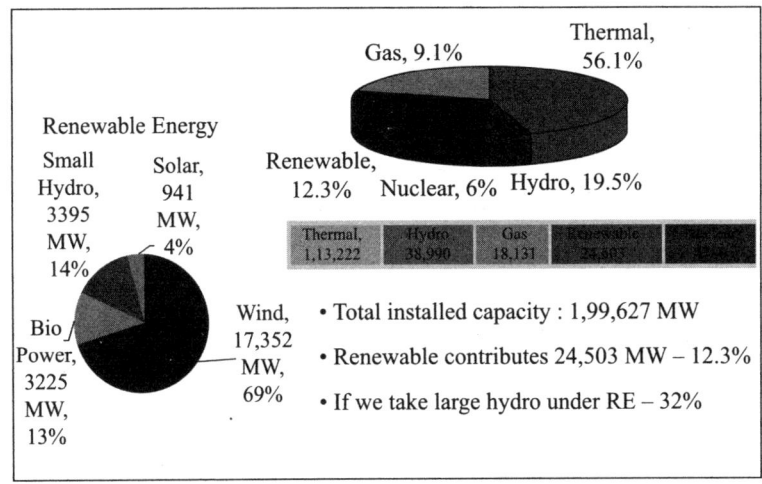

Source: Ministry of New and Renewable Energy (MNRE), Government of India, 2011.

Table 7.8

Renewable Power Potential

S. No.	Resource	Estimated Potential (In MW$_{eq}$)
1.	Wind Power	49,000
2.	Small Hydro Power (up to 25 MW)	15,000
3.	Bio-Power: Agro-Residues Cogeneration - Bagasse	17,000 5,000
	Waste to Energy: – Municipal Solid Waste to Energy – Industrial Waste to Energy	2,600 1,280
	Sub-Total	89,880
4.	Solar Energy	>100,000 30–50 MW/sq. km.
	Total	>1,89,880

Source: Ministry of New & Renewable Energy, Government of India, 2011.

Table 7.9

Plan-wise Renewable Power Growth

	Beginning of 10th Plan (MW) 1.4.2002	Beginning of 11th Plan (MW) 1.4.2007	Target 11th Plan (MW)	11th Plan Achvmnt. (MW)	Cumulative Achvmnt. up to 31.3.2012 (MW)
Wind	1,628	7,092	9,000	10,260	17,352
Small Hydro	1,434	1,976	1,400	1,419	3,395
Bio power	389	1,184	1,780	2041	3,225
Solar	2	3	200	939	941
Total	3,453	10,255	12,380	14,660	24,914

Source: Ministry of New & Renewable Energy, Government of India, 2011.

11 per cent in 2010 would increase to 18 per cent. This is more than the Chinese target of 15 per cent.

Wind Power

The country today faces an energy demand-supply gap of about 9 per cent with peak shortages being of the order of 10 per cent. The solution to this challenge lies in maximizing the utilization of renewable energy sources for meeting our energy demands. In that context, harnessing wind energy—the main grid connected renewable source of power—has become an indisputable reality for economic development. Wind energy has emerged as a viable, cost-effective commercial option for grid-connected power generation and as a competitive option to fossil fuel-based power generation. It is also the solution to deal with the ever-increasing prices of fossil fuels and the growing concern over global warming due to emissions from fossil fuel-based power plants.

The multidimensional initiatives which the Ministry of New and Renewable Energy (MNRE) has taken over the last three decades under its Wind Power Programme has been able to achieve large-scale commercialization of cost-effective generation of grid-quality wind power. That programme includes: comprehensive wind resource assessment;

broad-based research and development activities; implementation of demonstration projects to create awareness; development of infrastructure; capacity for the design, manufacturing, installation, operation and maintenance of wind turbines and a conducive policy environment involving many promotional measures at both the central and state levels. A key aspect of those measures is to encourage private investors and developers to take up commercial projects.

Table 7.10 indicates today's global scenario i.e., the ranking of countries on the basis of their installed wind power generating capacity as of end 2011. It will be seen that India is the 5th largest wind power producer globally after China, the US, Germany and Spain. However, we have to go a long way if we are to harness the 50,000 MW of wind energy-based power potential available on shore. As for the off-shore potential, it is practically limitless. This is, because of two reasons: first, we have a 7,500 kilometre-long coastline with good wind energy sources and profiles in many areas; and second, companies like Suzlon Energy Ltd (SEL), Pune have already acquired the basic technology for commercial

Table 7.10

Global Scenario: Total Global Installed Capacity as on December 2011

Country	Installed Capacity (MW)
China	62,364
USA	46,919
Germany	29,060
Spain	21,674
India	16,084
France	6,800
Italy	6,737
UK	6,540
Canada	5,265
Portugal	4,083
Denmark	3,871
Rest of world	28,272

Source: Global Wind Energy Council.

scale off-shore wind turbine manufacture, installation, commissioning, operation and maintenance.

Wind Resource Assessment

The wind climatology in our country is mainly governed by monsoon circulations. During the strong south-west summer monsoon, which starts in May–June, cool, humid air from the ocean moves towards the land and in the weaker north-east winter monsoon, which starts in October, cool, dry air moves towards the ocean. Wind energy is intermittent and highly site-specific. Therefore, a comprehensive wind resource assessment programme is essential for deciding the optimal sites for wind farms. Consequently, the emphasis was, and continues to be placed on wind climatology from the beginning of the Wind Energy Programme in the late 1980s. As a result, today, India has abundant data, collected from about 650 wind monitoring masts in 31 states and union territories (UTs).

The Wind Resource Assessment Programme is being implemented by the Centre for Wind Energy Technology (C-WET), an autonomous S&T institution of the MNRE as well as by the State Nodal Agencies for Renewable Energy. C-WET also assists developers in determining areas most suitable for deployment of wind power systems through wind maps of the country. Seven volumes of the handbook 'Wind Energy Resource Survey in India' have been published so far, covering wind data for 234 sites in 13 states/UTs. As a result, the wind power potential has been currently estimated to be around 50,000 MW assuming 2 per cent of the land availability for wind power generation in the potential areas. However, a major part of the potential (around 70 per cent) lies in low-wind regimes having wind power density of only 200–300 watts per square metre.

Demonstration Wind Power Projects

Demonstration wind power projects were taken up by the Ministry in the mid-1980s with the basic objective of creating the necessary infrastructure to open up sites for commercial development by demonstrating techno-commercial success so that the private sector gets the confidence to invest in commercial projects at those sites. The first demonstration project was set up near Tuticorin in Tamil Nadu in 1986. Subsequently, three wind farms with a total capacity of 20 MW were set up at Kayathar (6 MW), Muppandal (4 MW) in Tamil Nadu and Lamba (10 MW) in Gujarat.

With an aggregate demonstration project capacity of 71 MW established at 33 locations in nine states viz., Andhra Pradesh, Gujarat, Karnataka, Kerala, Madhya Pradesh, Maharashtra, Rajasthan, Tamil Nadu and West Bengal, most of the potential sites have been demonstrated and it has led to rapid commercial development. These projects have helped private investors in analyzing the performance and economics of wind power projects and identifying areas of concern in their operation and measures to overcome those concerns.

Commercial Development

Technology Development and Manufacturing Base

Wind Electric Generator Technology has evolved very rapidly in the country. State-of-the-art technologies are now available for the manufacture of wind turbines. All the major global players in the field have their presence in the country. The unit size of machines has gone up from 55–100 KW in the 1980s, to 3.0 to 3.5 MW today. Forty-four different models of wind turbines are being manufactured by 18 manufacturers, through (a) joint ventures under licensed production (b) subsidiaries of foreign companies, and (c) Indian companies with their own technology. Indigenization level of up to 80 per cent has been achieved. The current annual production capacity of domestic wind turbines is about 3,000 MW, which can be easily expanded to 5,000 MW, if the market demands.

The technology is moving towards better aerodynamic design; use of lighter and larger blades; higher towers; direct drive; and variable speed gearless operation using advanced power electronics. The technology is being continuously upgraded, keeping global developments in view. The industry has taken up indigenized production of blades, generators, yaw components, gear boxes and other critical components. Wind turbines and wind turbine components are being exported to the US, Australia, Europe, Brazil and other Asian countries. Due to India having a strong domestic manufacturing capacity and capability, some companies are now sourcing more than 80 per cent of their wind turbine components from India. This has resulted both in more cost-effective production at home due to larger production volumes, and in creating additional local employment.

Status of Indian Technology of Wind Turbines vis-à-vis Global Levels

Table 7.11 compares Indian-made and leading edge internationally-made wind turbines in terms of four key technological parameters.

It will be seen that technologically, Indian-made wind turbines are practically at the leading edge of technology at the global level. Indeed, the Indian turbine industry had, by 2010, practically closed the technology gap that existed in 1996.

Table 7.11

Comparison between Indian-made and Internationally-made Wind Turbines

Sr No	Parameter	Indian	Leading Edge Globally
1.	Rating/Capacity	250 KW (1996) 3.0–3.5 MW (2010)	4.5 MW Typical (2010)
2.	Hub Heights	41m (1996) 90 m (2010)	100 m Typical (2010)
3.	Rotor Diameter	28 m (1996) 100 m (2010)	120 m Typical (2010)
4.	Whether Geared or Gearless Turbines	Both	Both

Source: Ministry of New & Renewable Energy, Government of India, 2010.

Suzlon Energy Ltd (SEL)

Suzlon Energy Ltd (SEL), established in 1995, is by far the largest wind turbine manufacturer in India. Over 1997–2011, SEL has consistently had a market share of more than 40 per cent. Its cumulative installed base of turbines in India as of the financial year 2011–12 was 7,357 MW. Figure 7.3 shows SEL's Annual Capacity Addition in MW over 1995–96 to 2011–12. SEL has over 1,600 customers and over 40 wind farms across eight states.

Crowning these achievements is the fact that SEL is constructing a 1,500 MW Wind Park in Kutch, Gujarat to be completed by 2015, 950 MW of which has already been installed and operationalized at end-2011–12. Currently, the largest wind farm in the world, its construction period would be a record four years only. This Wind Park uses 690 of SEL's own turbines with power ratings of 600 KW, 1,250 KW, 1,500 KW & 2,100 KW and involves a total capital investment of ₹ 5,700 crores (approximately US$ 11 million). With the huge scale economies involved, the cost of power produced is expected to be 30 per cent less than normal levels, the lowest in the world.

In 2007, SEL totally bought out the German wind turbine manufacturer, the RE Power Ltd. The RE Power-amalgamated-SEL has a product range covering all customer needs of turbines from 0.6 MW to 6.0 MW including pioneering offshore turbines of 6.15 MW capacity, again the largest in the world and far ahead of China. The buy-out has created the world's fifth largest wind turbine manufacturer with a global market share of 8.4 per cent, a market capitalization of US$ 4.0 billion and a total annual turbine manufacturing capacity of 7,500 MW. SEL-RE Power (The Suzlon Group) has a workforce of 13,000 in 32 countries serving 1,600 customers in six continents. Truly a giant Indian Trans-National Corporation. No Chinese wind turbine manufacturer comes anywhere near these performance levels.

The global installations of the Suzlon Group by continent and country as of end of 2011–12 are shown in Table 7.12.

Centre for Wind Energy Technology

The Centre for Wind Energy Technology (CWET) , an autonomous S&T institution of the Ministry of New and Renewable Energy (MNRE) was established at Chennai in 1996. It serves as the S&T focal point for wind power development in the country. The National Wind Turbine Test

Figure 7.3

Annual Capacity Addition of SEL (1995–96 to 2011–12)

- Suzlon is No.1 for 13 years in india with > 40% market share
- Cumulative installed base in India of more than 6200 MW
- 1500 customers and over 40 wind farms across 8 states
- Developing one of world's largest wind parks at Kutch, Gujarat. Already over 800 MW installed in Kutch with planned capacity of 1500 MW

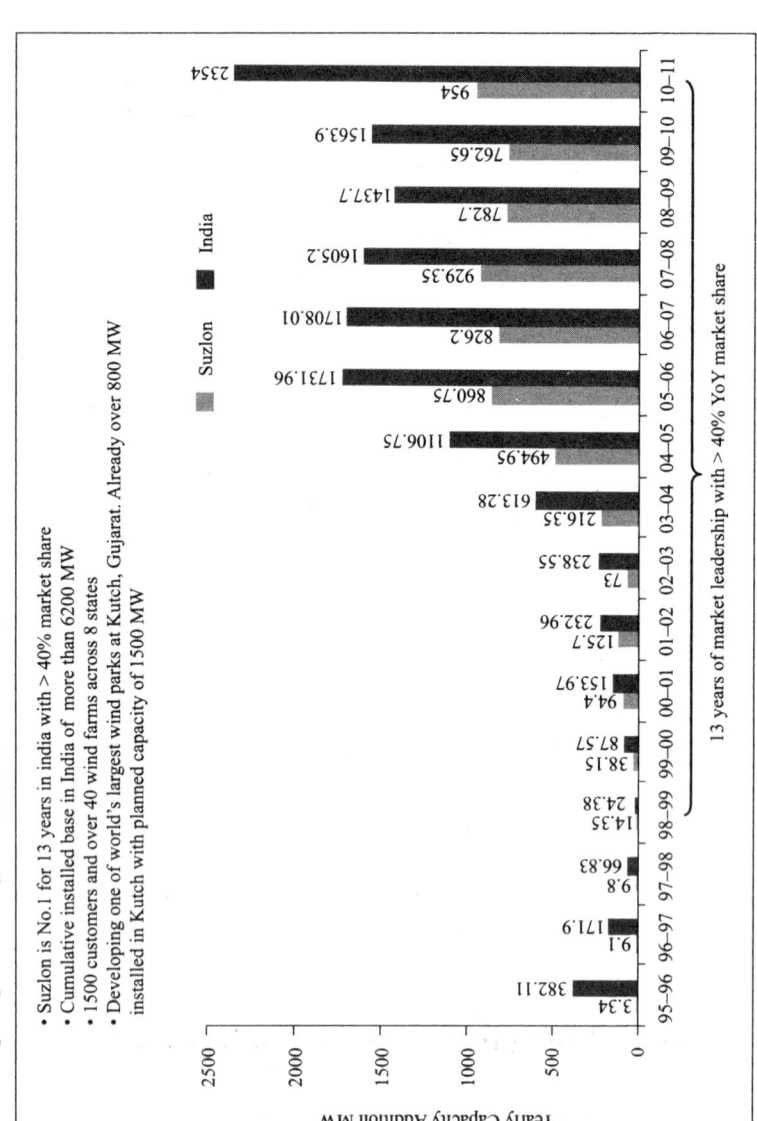

Table 7.12

Global Installations of the Suzlon Group (as of end of 2011–12)

Asia and Australia	
Country	Number
India	7379
China	1162
Australia	912
Japan	118
Sri Lanka	10
Total	9581

Latin America	
Country	Number
Brazil	388
Nicaragua	63
Total	451

Europe	
Country	Number
Germany	2196
France	1303
UK	795
Italy	646
Others	958
Total	5898

North America	
Country	Number
USA	3195
Canada	45
Total	3240

Source: Author.

Station of C-WET has been established at Kayathar, Tamil Nadu, with technical support from RISO of Denmark.

Objectives of C-WET

The main tasks of C-WET are to:

- Coordinate and support research and development programmes to achieve and maintain reliable and cost-effective technology in wind power systems.
- Analyse and assess wind resources and prepare wind energy density maps, wind atlases and reference wind data.
- Prepare and establish standards, and undertake testing and certification of wind power systems in line with internationally recommended practices and standards.
- Conduct and coordinate the testing and evaluation of complete wind power systems.
- Accord Type Approval/Type Certification, for wind turbines.
- Monitor the field performance of wind power systems, and serve as the national information centre for selective dissemination.
- Undertake human resource development programmes in the wind energy sector.

C-WET has the following units to carry out its various activities:

- Wind Resource Assessment Unit
- R&D Unit
- Standards & Certification Unit
- Testing Unit
- Information and Training Unit

Research and Development (R&D) Unit

The R&D Unit supports time-bound and mission-oriented research and development programmes to achieve and maintain world class, reliable and cost-effective technology in wind power systems with a mutually beneficial inter-disciplinary approach for most of its projects. The strategic collaboration that could assist in most technological development for our country is nurtured by funding a network of institutions and providing needed technical support. The major research projects in progress and executed are detailed as follows:

Generator and Grid Integration of Wind Turbines

Study on Power Quality Issues in Grid-Connected Wind Farms and Identification of Remedial Measures

The wind power industry is currently facing serious power quality issues, as most of the wind turbines across the country are installed in remote/rural locations where the grid network is rather weak. To overcome these issues faced by the wind energy stakeholders/developers, C-WET has initiated a project to identify the power quality issues of wind farms and the remedial measures essential for improving the grid behaviour of the existing wind turbines.

Power Evacuation Studies for Grid Integrated Wind Energy Conversion Systems

The second issue faced by the wind turbine industry/stakeholders is high wind generation and low demand. Consequently, the wind turbine systems often have to be shut down for want of proper power evacuation infrastructure. C-WET is undertaking this R&D project to identify how to facilitate efficient evacuation of wind power. The studies done so far have indicated good progress in scenario analysis and is expected to help in optimizing wind turbine operation by ensuring evacuation of available wind power to the grid with high penetration.

Wind Turbine Blades

Experimental Characteristics of Wind Turbine Blades over the Full 0–360 Degree Angle of Attack

There is a dire need for optimization of aerofoils/wind turbine blades so that they are effective in energy extraction in the low and moderate wind regimes and dusty environments prevailing over most of the country. C-WET has therefore initiated a project, at C-WET, to understand aerofoil characteristics (0–360 degree angle of attack) and detailed aerodynamic behaviour of flow around aerofoils in the regions immediately before and after stall to understand the stall hysteresis mechanism.

R&D Infrastructure Up-gradation

Dedication of a 2 MW R&D/Experimental Wind Turbine to the Country at WTRS Facility, Kayathar

A 2 MW experimental/research wind turbine facility has been successfully commissioned by C-WET at its Kayathar Wind Turbine Research Station. The instrumentation on this state-of-the-art variable speed wind turbine will provide very useful insights to design, operate and monitor wind turbines.

Human Resource Development (HRD)

HRD is one of the key problem areas facing our wind power industry; viz., in getting trained engineers/qualified technical personnel particularly for the operation and maintenance cost (O&M) of wind turbines for their expected design life of 20 years. As part of R&D on HRD, C-WET has initiated courses on wind power S&T, in collaboration with the Indian Wind Turbine Manufacturers Associations, in PSG College of Technology, Coimbatore and Amrita School of Engineering, Coimbatore.

Wind Resource Assessment (WRA)

WRA in Uncovered/New Areas

The National Wind Resource Assessment (WRA) Programme has been implemented by C-WET in association with the state nodal agencies since 2003. At present, India has the largest wind resource assessment programme in the world; and wind monitoring stations for resource assessment have been set up covering the entire length and breadth of the country (28 states and 3 union territories). So far, 653 wind monitoring stations have been established under the Ministry's WRA programme.

Wind Measurement at 120 Metre Level

Four guyed towers of 120 metre height for anemometry are operational at Lamba (Gujarat), Akal (Rajasthan), Jagmin (Maharashtra) and Jogimatti (Karnataka). Measurements have been done at 10, 30, 60, 90 and 120 metres height. The data from these stations would serve as valuable research material for wind energy profiling and related applications.

Consultancy Projects

C-WET has carried out WRA-value added services through more than 14 projects in various areas like Micrositing & Technical Due Diligence, Installation of Wind Monitoring Stations, Power Performance Guarantee Test etc., over 2005–10. It has also prepared a detailed project report for the establishment of a 10 MW wind farm project at Bidda in the Reasi district in Jammu and Kashmir for National Hydroelectric Power Corporation and the J&K Renewable Energy Development Agency. The report has been accepted by the clients and ground work is due to start soon.

Estimated Wind Power Potential in India

As already mentioned, as per the Indian Wind Atlas, the installable wind power potential has been estimated as 50 GW at 50 metres above ground level with the assumption of 2 per cent land availability in the wind potential area. As the current technology brings up wind turbines of higher hub heights in the range of 80–120 metres, and as land use pattern had not been considered in previous assessments, it is necessary to know the potential at higher hub heights with the details of actual land availability. The potential can be described on the basis of different scenarios. The 'moderate' scenario takes into account all existing or planned policy measures and also the advantages of modern wind turbines in terms of new hub heights, rotor diameters and overall efficiency. Details are given in Figure 7.4. In this context, a similar exercise has been carried out for estimating potential at 80 metre level. However, the actual wind resource available and land availability have to be assessed and validated with the actual meteorological mast-based measurements and land surveys respectively.

Figure 7.4

Wind Power Potential in India

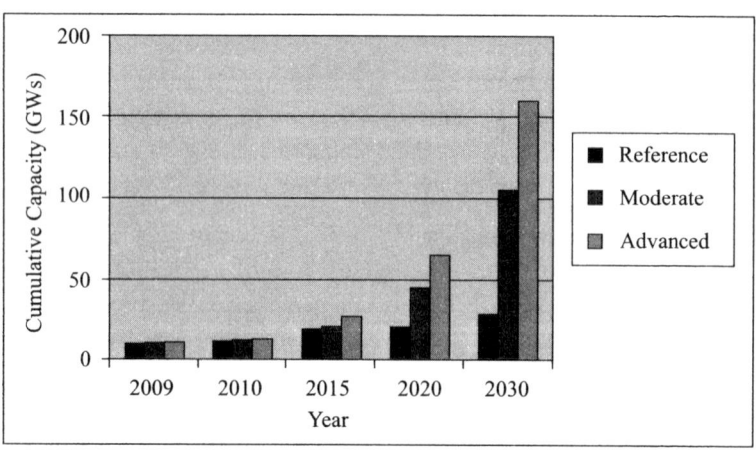

Source: Global Wind Energy Council.

Wind Resource Assessment at 100 Metre Level

A project has been initiated by C-WET to conduct a realistic assessment considering tangible land availability for wind farming, for seven wind potential states, namely, Tamilnadu, Karnataka, Andhra Pradesh, Maharashtra, Madhya Pradesh, Gujarat and Rajasthan at 100 metre level thereby facilitating validation of meso-scale based results indicated in the Wind Atlas. Launched in 2011, the project duration is three years and would be funded by the Ministry of New and Renewable Energy.

Development of Wind Forecasting Model with Special Reference to Complex Terrain

A project titled 'Development of a Wind Forecasting Model with Special Reference to Complex Terrain' has been sanctioned by the Ministry. C-WET approached RISO, Denmark and they have agreed to carry out the project in collaboration with C-WET. An agreement was executed in June 2010 between C-WET and RISO to execute the project. Project duration is of 18 months. The project cost of ₹ 20 lakh is to be shared by MNRE and C-WET. The 'PREDICTOR' model has been identified and the same is being used for the forecasting. A wind farm in complex terrain with a 50.4 MW capacity has been identified for the study at Khandke Hills, Ahmednagar, Maharashtra. RISO conducted a workshop on 'Wind Power Forecasting' at C-WET. The project is ongoing.

Offshore Wind Resource Assessment at Dhanuskodi, Rameshwaram, Ramanathapuram District, Tamil Nadu

The Ministry of New and Renewable Energy has sanctioned a project to study the wind resource at Rameswaram by installing a tall tower (100 metres) at Dhanuskodi with the objective to examine the feasibility for setting up an offshore wind farm project there and in and around the long stretch extending towards Sri Lanka. In the meantime, the data captured above the sea surface by a Synthetic Aperture Radar (SAR) obtained from the European Space Agency and converted into 10 metre height wind speed data. Based on these inputs, CWET has prepared a report in association with RISO. RISO have suggested off-shore wind measurements at five locations to validate the results of the SAR study. This report also indicates good offshore potential in the Kanyakumari–Rameswaram belt. Further, INCOIS, Hyderabad has also studied the offshore resource with reference to QuickSCAT (NASA, US) data and prepared a report. Finally, MNRE/CWET approached Scottish experts through the Scottish Development International in November 2010 to get their technical opinion about the preliminary offshore wind estimate. The S&T work is ongoing.

Wind Turbine Testing

C-WET's Wind Turbine Test Station (WTTS) near Kayathar in Tamil Nadu was established with the technical assistance of the RISO National Laboratory, Denmark with technical assistance and partial funding from the Danish International Development Agency (DANIDA) and with the remaining financial assistance and guidance from the Ministry. The test station has the following facilities:

- Two test beds to test wind turbines up to (i) 1,250 KW capacity and (ii) 850 kW capacity, which capacities are expandable based on requests from potential customers.
- Readily available grid connection for each test bed.
- Readily available reference meteorological masts in front of each test bed, designed for heights of 75 metres and 50 metres respectively.
- Industrial PC-based Data Acquisition Systems for measurements at the control room of each test bed.
- An office cum workshop building at WTTS with facilities for carrying out functionality check of instruments and sensors. The workshop is equipped with adequate space to accommodate a nacelle for instrumentation purposes.

- Sensors and transducers as per the requirement of IEC standards stored in the workshop as per the International Quality Management System Procedures.
- Nine wind turbine generators (WTGs) of 200 kW Micon make, for development and testing of new innovative measurement techniques.

Standards and Certification

C-WET has completed the renewal of Provisional Type Certificates of three wind turbine models. The formulation of Indian Standards on wind turbines is in progress, in close co-ordination with the Bureau of Indian Standards. C-WET, as a technical body, has successfully completed 10 years in review of documentation/information provided by various wind turbine manufacturers, in connection with issuing a revised list of models and manufacturers of wind turbines.

Preliminary exposure was provided to C-WET scientists on NREL Codes at the National Renewable Energy Laboratory (NREL), US. The structural model of a wind turbine has been created using those codes as well as commercially available and industry-recognized aero-elastic code, by C-WET scientists.

Information Training and Commercial Services

National Training Courses

C-WET has successfully organized 10 national training courses on 'Wind Energy Technology' with the objective of providing basic knowledge on wind turbine technology and also a platform to exchange views and experiences with wind energy experts. The programme has provided comprehensive knowledge right from wind resource assessment to installation and commissioning of wind farms along with technical and financial challenges.

International Training Courses

C-WET had successfully organized the 7th International Training Course on 'Wind Turbine Technology & Applications'.

Infrastructure Management

The centre manages and maintains the central server, e-mail, internet, intranet, e-security and computers inside the campus. The Professor Anna Mani Information Centre, a part of the C-WET library and managed by C-WET is currently being extensively updated. It consists of huge knowledge resources on wind energy and other renewable energy sources. The unit also maintains a display hall, training infrastructure (seminar hall, conference hall, audio-visuals, etc.).

Deployment

The successful operation of the demonstration projects attracted a number of large investors towards wind. The first wind farm in the private sector, of 3 MW capacity, was set up in Tamil Nadu in 1992 and the energy was wheeled to the developer's industry for captive use. Since then, other states, like Gujarat and Andhra Pradesh, have adopted the TNEB model. Overall, the annual addition in wind turbine generating capacity picked up from the start in 2002, and the bulk of that addition (7,040 MW) has been in Tamil Nadu. The total wind power capacity reached 17,500 MW as of April, 2012, the main states involved being Tamil Nadu, Gujarat, Maharashtra, Andhra Pradesh, Karnataka and Rajasthan. The year 2011–12 was a good one for the Indian wind power industry. The capacity addition was 3,200 MW. That was the first time that total annual installations crossed 3,000 MW. The wind power capacity contributes about 8 per cent of the total grid-connected installed power-generating capacity, and it is around 70 per cent of the total renewable power capacity installed in India.

Guidelines for Setting Up Projects

The Ministry of New and Renewable Energy has been issuing comprehensive guidelines for wind power projects to all stakeholders from as far back as July 1995 to bring about healthy, orderly and rapid growth of the wind energy sector and to achieve optimum generation of power in the most efficient and cost-effective manner. These guidelines relate to: preparation of detailed project reports (DPRs), micrositing, selection of wind turbine equipment, operation and maintenance, performance evaluation, etc. They have created awareness in the state electricity boards, state nodal agencies, manufacturers, developers and investors about planned development and implementation of wind power projects. The certification requirement for wind turbines was re-introduced with time-bound provisions for self-certification in 1996. The C-WET issues a list of manufacturers of certified wind turbines on a quarterly basis under different categories. The guidelines have helped in the installation of quality wind turbines and orderly growth in the country.

Fiscal Incentives and Promotional Policies

The Government promotes wind power through fiscal incentives such as concessional customs duty on certain components of wind electric generators, total excise duty exemption, a 10-year tax holiday on income generated from wind power projects, and loans from the Indian Renewable Energy Development Agency (IREDA) of MNRE and other financial institutions. MNRE has been issuing guidelines to all state governments to create an attractive environment for the purchase, wheeling and banking of electricity generated from wind power projects. Incentives were given to encourage industrial companies and businesses to invest in wind power. Preferential tariff is being provided to increase wind energy generation in the states having wind power potential. Tamil Nadu, Andhra Pradesh, Haryana, Karnataka, Madhya Pradesh, Rajasthan, Maharashtra, Gujarat, Punjab, Kerala and West Bengal have announced to give wind power preferential tariffs ranging from ₹ 3.40 to 5.00 per kWh. A total of 15 State Regulatory Commissions have announced the Renewable Power Obligations (RPOs), which mandate the Distribution Licensees to necessarily take a certain percentage of electricity from renewables, which has accelerated the growth. Government is trying to get the remaining states to also announce the RPOs. Efforts are also on to harmonize the RPO percentage in various states and enforce them.

New Initiatives

Wind Atlas

A national numerical Wind Atlas has been prepared and launched. It will help all stakeholders to achieve better micrositing leading to higher generation from wind power projects. The Wind Atlas has been prepared by C-WET in association with the RISO National Laboratory of Denmark. RISO is the pioneer institution for the development of Wind Atlas for many countries. The modelling techniques used for preparation of Wind Atlas for India include Wind Atlas Analysis & Application Programme (WAsP) and Karlsruls Atmospheric Mesoscale Model (KAMM). The vast amount of field wind data being collected on a continuing basis under the MNRE's wind monitoring programme is also being used for validation of the Atlas.

Generation-based Incentives Scheme

The main driving force and the real incentive for development of wind power has been the provision for accelerated depreciation of 80 per cent. This provision has enabled large profit-making companies, small investors and captive users to participate in the sector. However, in the last few years, the power scenario in India has witnessed a qualitative change with the entry of Independent Power Producers (IPPs) and Foreign Direct Investors, who cannot avail of the accelerated depreciation provision. Therefore, to broaden the investor base, and encourage better generation efficiencies, MNRE introduced in December 2009 a Generation-Based Incentives (GBIs) Scheme of 50 paise per unit for such classes of investors. The response to the scheme has been modest. However, despite such response, 1,990 MW of wind power projects have been installed under the scheme. The scheme was meant to achieve a target of 4,000 MW over the 11th Plan (2007–11). However, in view of demands from various stakeholders to continue the GBI benefit, the MNRE is examining the matter for further action.

Preferential Tariff Rates

According to the Electricity Act 2003, state utilities are to encourage wind energy development by providing preferential tariff rates for generation of electricity from all renewable energy sources including wind energy. The MNRE has been constantly interacting with various State Electricity Regulatory Commissions (SERCs) and the Central Electricity Regulatory Commission (CERC) to finalize the preferential tariff policy for wind power projects. CERC introduced the guidelines for determining the tariff. In 2009, states like Maharashtra and Rajasthan have issued the tariff policy as per the CERC guidelines. However, the remaining states have yet to revise their policies.

Renewable Energy Certificates

A new system of Renewable Energy Certificates (RECs) has been introduced in the country, by which energy generated and fed to the grid will entitle the developers to trade the RECs issued to them. The system has to be established with time. This will further enhance the growth.

Global Scenario

Wind energy has emerged into a mature and booming global business. Generation costs have fallen over the last few years, moving competitively with conventional energy sources. The technology has improved dramatically in view of higher capacity machines, higher hub-heights, efficient and reliable technology. In view of the concerns shown for global warming, there is a huge and growing global demand for the environment-friendly wind power. Over the last decade, global wind installations have continued to grow at an average growth rate of around 30 per cent. The world witnessed the year 2011 to be another record year for the wind power development with over 43 GW of new installations, taking the cumulative installed capacity to reach 2,37,669 MW.

In 2011, although China remained the leader. India ranked third in terms of new installations, after China and the US. Major contributing countries are China (62.36 GW), United States (46.92 GW), Germany (29.06 GW), Spain (21.67 GW) and India (16.08 GW) upto December 2011.

Future Perspective

The target of 9,000 MW set for the 11th Plan has been over achieved by installing 10,260 MW. The MNRE has set a target for additional capacity creation during the twelfth Five-Year Plan (2012–17) of 15,000 MW. Figure 7.4 is a longer term forecast upto 2030. The target set for the first year of the plan viz., 2012–13 is 2,500 MW. We need to look forward to some policy reforms and institutional improvements. Though C-WET has done well on the exploitation of the national wind resource, the potential needs to be reassessed both in extent and intensity. The industry and developers are looking forward to the GBI scheme to continue in the 12th Plan also. The future will depend upon the impact of the removal of the over two-decade old accelerated depreciation provision for this sector, consequent to the introduction of the Direct Tax Code (DTC) and the Goods and Service Tax (GST). The large public sector entities such as the railways, the telecommunications companies, oil companies and the public sector 'Navaratna' companies, particularly those that are big consumers of power themselves, need to go in for wind power generation. Wind power projects are capital-incentive and therefore, the public financial institutions have to be motivated by the government providing even easier financing for them.

The Chinese Wind Power Industry

In the wind energy sector, China has taken the path of a 'fast follower' within a period as short as five years. Wind technology has been diffusing through China at a rapid pace and has had a cumulative average annual growth rate of more than 100 per cent over the last four years. In terms of cumulative wind energy installations China, in 2011, surpassed the US to take the global number one position.

This rapid rate of diffusion has also been accompanied by a steady increase in the relative size of installed wind turbines. However, the vast majority of Chinese wind turbines are still only of 1.5 MW rating in 2011. In contrast in India the vast bulk of the turbines made even in 2009 were of 3 MW rating. The Indian target for 2015 is 5 MW rating. Furthermore, China's export of wind turbines has been nil.

Driving Forces of Development

The driving forces for China according such very high priority to wind energy development are:

- Energy Security through Technology Leapfrogging.
- Environmental Pressure.

  Environmental pressure on China is growing as it is the world's biggest green house gases polluter, emitting 6 billion tons of energy-related CO_2, 55 per cent of which is due to power generation. Sixteen of the world's 20 most polluted cities are in China. The Chinese power sector is the key sector for emission reductions. Wind is the most promising source of clean energy for large grid connected power.
- Economic and Social Potential

  The Chinese wind market is forecast to see CAPEX of US$ 91 billion up to 2020. This CAPEX will be a major spur to full domestic supply chain development and huge employment opportunities.

Structure of the Innovation System

A comprehensive supply chain has developed in China. The five state-owned electric utilities also dominate the wind farm market and are the major buyers of wind turbines made in China. These wind power base projects make up the majority of new installations.

Mode of Project Execution

In China, most wind energy projects are set up on a turnkey basis by the wind turbine (WT) manufacturers for one or other of the nation's 10 giant electric power utility companies. They represent more than 75 per cent of installed capacity. The residual capacity is brought in by as many as 85 companies, 90 per cent of which are also state-owned.

Regulation as the Most Important Part of Institutions

The main government agencies in charge of renewable energy policies are the National Development and Reform Commission (NDRC) and its Energy Bureau and the National Energy Commission. Early key actions included the power sector reform in 2002, which aimed at facilitating access to the grid for renewable energy generation. China has clear targets for and a political commitment to renewable energy sources in general and wind power in particular. Major progress was made in 2006 when the Renewable Energy Law took effect. This law created a more stable and long-term perspective for renewable energy. The features of Chinese regulation policies can be grouped into policies affecting the supply side, on the one hand, and the demand side on the other.

The Ride Wind Programme (RWP), initiated in 1996 at the start of China's ninth Five-Year Plan (1996–2000), specified that the demand for wind farms would be created by the government, while wind power equipment production would be undertaken by joint ventures between Chinese and foreign firms, with the latter being offered the opportunity to participate in the projects in return for technology sales. The RWP also required, for the first time, the use of locally made components (40 per cent of the total WT value), which effectively introduced wind turbine manufacturing technology into China. During the Wind Power Concession Project period from 2003 to 2007, it was mandated to source at least 70 per cent of the total value from domestic manufacturers.

The local content requirement has thus been a very direct way to support the formation of a domestic wind energy industry. The local content requirement was officially abolished in December 2009. However, whether this will facilitate access to the mega projects for foreign players remains unclear. In any case, the local content requirement seems to have accomplished what it was designed to achieve viz., the creation, in a 20-year time horizon, of a strong domestic wind energy supply chain large enough to cope with China's rising demand for wind turbines.

The build-up of human resources is another important indirect form of support. In 2005, China's Education Ministry authorized the setting up of a four-year degree programme in wind energy at the North China Electric Power University (NCEPU) to meet the need for engineers and technicians for the wind energy sector: it has campuses both in Beijing and the nearby Hebei province. It is the only university in China dedicated to human resources development for the wind sector.

The Wind Power Concession Project sought to promote the establishment of large wind farms (each farm having a capacity of at least 100 MW) in which government-owned power grid companies sign power purchase agreements with wind power project investors with the term of the agreement covering the total forecast operational period.

There are also more indirect forms of supporting the build-up of a domestic wind turbine industry. One of them is procurement. In 2006, three Chinese ministries jointly released the Provisional Measures for the Accreditation of National Indigenous Innovation Products. These measures establish a procedure under which products made with 'indigenous' (for example, Chinese) intellectual property can qualify for priority in government procurement. Because it is proving very difficult for foreign enterprises to qualify for 'indigenous' status under this programme, the measure effectively led to procurement preferences favouring the domestic renewable equipment manufactures. Furthermore, the Ministry of Finance issued a regulation on import tax and customs duty exemption for wind turbine components. Only wind turbines with a capacity larger than 2.5 MW as well as some stand-alone key components (converters, bearings, controls) are eligible. The goal is to ease the supply bottleneck for components which are not yet produced in China.

The use of technical standards is also an important instrument being used to foster domestic industries. There are a number of internationally recognized standards in the field of renewable energy; for example, those issued by the International Organization for Standardization (ISO) or the International Electro-technical Commission (IEC). As many countries, such as Germany and the United States have done, China also released its own technical standards. Such standards are typically scrutinized from the viewpoint of whether or not they hinder international trade, and are often subject to WTO disputes. A WTO-related dispute involving wind energy arose after the United States challenged the Special Fund for Wind Power Equipment Manufacturing as the WTO prohibited subsidies. However, after that challenge, the US and Chinese governments resolved the dispute after several rounds of negotiations.

The main policies on the demand side are associated with the Renewable Energy Law, enacted in 2006, and the Medium and Long-Term Development Plan for Renewable Energy, released by the NDRC in 2007. The Renewable Energy Act established a framework under which utilities would be required to pay the full price for the electricity generated by renewable energy sources.

The Medium and Long-Term Development Plan for Renewable Energy put forward national renewable energy targets and priority sector policy measures for implementation. It also set for the first time a Mandatory Market Share (MMS) for electricity from renewable energy. The Plan called for a medium range target of 100 GW installed capacity in 2020. It required power companies which owned installed capacity of more than 5 GW to have non-hydro renewable energy installed power capacity accounting for 3 per cent of total capacity by 2010 and 8 per cent by 2020. This measure has triggered a surge of investment in wind energy, reflecting the fact that wind power equipment is less costly to install and operate than solar and biomass alternatives. In March 2011, the Chinese government updated the targets and called for a medium range target for 2020 of 180 GW installed capacity.

Feed-in-Tariff

In July 2009, China's NDRC introduced a feed-in tariff for wind power which applied for the entire operational period of a wind farm (20 years). There are four different categories of tariff depending on a region's wind resources. These are considerably higher than the tariff paid for coal-fired electricity.

'Three Gorges in the Air' Programme

In order to push forward the development of the wind energy sector, in August 2009, the National Energy Administration (NEA) started the construction of a series of 10 GW 'Three Gorges in the Air' wind farms in selected locations in the provinces with the best wind resources and set targets for each of them to be reached by 2020. So far, the Chinese government has confirmed seven giga watt (GW)-scale wind power bases. According to the plan, wind power bases will add upto 138 GW of wind power capacity by 2020, on the assumption that a supporting grid network is established.

Removal of Local Content Requirements

In January 2010, the Chinese state media reported that the NDRC had 'scrapped' the 70 per cent local content requirement. While the stipulation for local content for government promoted wind farms may have been removed, China did not eliminate Article 25 (5) of the Renewable Energy Law, which provides that the Renewable Energy Development Fund be used to promote "localized production" of equipment for the development and utilization of renewable energy.

Offshore Wind Power Generation

The China Meteorological Administration has estimated China's offshore wind potential at more than 750 GW—far higher than the 250 GW potential for land-based wind. China's eastern coastal areas boast sound conditions to develop wind farms on beaches and in offshore areas. The first stage of construction of the East Sea Bridge Offshore Wind Farm—China's first large-scale offshore wind farm located in Shanghai—was completed in June 2010 and successfully connected to the Shanghai grid by a submarine cable in July 2010. It has a total installed capacity of 102 MW and an estimated average annual power output of 268 gigawatt-hours (GWh). China expanded its installed offshore wind capacity from 63 MW in June 2009 to 142 MW in June 2011, with an average annual growth rate of 63 per cent for that period. According to the Chinese Renewable Energy Industries Association (CREIA), China will expand its offshore wind power installed capacity to 5 GW by 2015 and 30 GW by 2020.

Interim Measures for Offshore Wind Farm Development

On 23 January 2010, China's National Energy Administration (NEA) and State Oceanic Administration (SOA) jointly put into effect an Interim Measure on the Management of Offshore Wind Farm Development. The Measures stipulated that all offshore wind projects will be subject to the approval of the NEA. The provincial-level Energy Administration Department is responsible for drawing up plans for local offshore wind energy development, whilst the provincial-level Oceanic Administration Department is to provide preliminary opinions on the use of sea space and on a project's impact on oceanic environments. Developing and investing companies should be Chinese-funded companies or Chinese-controlled joint venture companies (in which the Chinese party holds at least a 50 per cent stake).

Grid Connection Problems

The rapid development of wind power in China has put unprecedented strain on the country's electricity grid infrastructure. In February 2011, China's State Electricity Regulatory Commission said that about 2.78 billion kilowatt-hours of power generated by China's wind farms was wasted in the first half of 2010, because of the lack of grid connectivity. Many local grids lack the capacity to absorb wind-generated electricity.

Added to this, the faster-than-expected growth in installation in the last few years has created additional bottlenecks. The output of some wind farms has been rejected by the local grid companies because of concerns that it would overload their systems. There are also limitations in the grid's capacity to transmit power over thousands of kilometres. For power distributors there is little incentive to invest in extra gridlines to remote wind farms because, while Beijing has subsidized wind power generation, the grid operators are getting the same tariff as power generated from coal.

Improving the Grid Situation

According to some analysts, the ground situation is not all that bad considering the methodology China uses for calculating the installed capacity. The Chinese Federation of Power Generation, which provides China's energy statistics, only counts wind farms as operational from the moment that the last turbine of a project has become grid-connected. However, in reality, most of the installed wind turbines of a project are connected to the grid and start generating power much earlier. Moreover, due to a lack of incentives, Chinese grid companies have been reluctant to accept large amounts of wind power into their systems. However, grid companies have recently reached an agreement to connect 80 GW of wind power by 2015 and 150 GW by 2020. At the end of 2010, China's State Grid invested RMB 40 billion (US$ 6 billion) to facilitate wind power integration into the national power grid. There are nine provinces in China that already have at least 1,000 MW of installed wind power. Through the end of March 2011, the State Grid Corporation had connected more than 33,000 MW of wind power to their grid (or nearly 75 per cent of installed capacity).

Wind Power Generation Equipment Manufacturing Industry

By 1999, roughly 40 Chinese companies were producing wind turbines, for the most part small 100 watt units used in single households.

Beginning 2005, China started planning to build of 1 GW-scale wind power bases. The planning and construction of 10 GW-scale wind power bases started in 2008. This strongly boosted the scale development of the wind power industry in China, created good market conditions for the domestic manufacture of wind turbines, stimulated rapid development of the turbine manufacturing industry and advanced the formation of independent companies. While there are foreign-owned wind energy equipment suppliers, the government's nationalistic stance has given huge, all-embracing advantages to domestic suppliers. In 2009, 77 per cent of the market was served by Chinese suppliers. The first tier of domestic suppliers is led by Sinovel, Goldwind, Dong Fang Electric and United Power (which are also among the world's top 10 WT manufacturers) followed by two foreign companies, Vestas of Denmark and Games of Spain in that order. All the Chinese companies started to develop their WT production based on technology purchased from foreign firms. However, China's dominant position today in the clean energy sector, particularly wind power, is due to it having followed a mix of state-directed regulation and systematic assimilation of foreign technology.

A Profile of the Leading Wind Power Companies

In 1986, the Inner Mongolia Autonomous Region (IMAR) government offered the first known subsidy for the development of local wind or solar energy equipment, a financial incentive of RMB 2000 (US$ 285) per wind turbine or PV panel and RMB 50 (US$ 7.5) per battery to consumers who bought equipment manufactured in the IMAR. Other regional governments offered similar incentives, which increased demand for local production and proved "successful in creating a domestic industry of wind turbine manufacturers". Sinovel was founded in 2004 as a spin-off of the giant Dalian Heavy Industry Corporation. It entered the wind power market through a technology transfer license for 1.5 MW WTs from the German company Fuhrlander. In line with the market trend towards larger WTs and the prospect of China's offshore potential, Sinovel added 3 MW turbines to its portfolio through a joint technology development with the US company, AMSC Wind Tec. That development is now in the process of being extended to 5 MW WTs also jointly with AMSC. Sinovel wanted to tackle the quality and reliability issues of its WTs, and it contracted another German company, Germanis Char-Lloyd, for certification of its turbines. Sinovel currently has a Chinese market share of 22 per cent.

The Xinjiang Goldwind S&T Company Ltd is a subsidiary of the Xinjiang Wind Energy Company (XWEC) and was established in 1998. Its parent organization, the Xinjiang Wind Energy Research Institute, a research-oriented industrial group, was founded a decade earlier in 1988. Substantial support was provided by foreign government programmes e.g. RISO of Denmark for the first wind related R&D and pilot plant activities of the group (XWERI). Technology has been acquired through license (TOT) agreements with various German companies (RE Power, Jacobs and Vensys). Following the emergence of Chinese supplier networks, Goldwind has been able to continuously increase the local content of its turbines. In early 2008, it acquired 70 per cent of Vensys, intensifying its already close collaboration with that company on large-scale direct drive turbines.

Dongfang Wind Power was established as a strategic diversification of the Dongfang Electric Corporation, one of the world's largest power generation equipment manufactures. Its product portfolio includes a 1.5 MW turbine made under license from the RE Power of Germany and a 2.5 MW WT jointly developed with the German Aerodyne.

Too Many Local Players

According to the Chinese Renewable Energy Industries Association, there were 86 enterprises engaged in manufacturing complete grid-connected wind turbines at the end of 2009 in China. With over 80 wind turbine manufacturers in China, total production capacity for 2011 reached 29 GW, far exceeding the actual demand of 15 to 18 GW. This trend is evidence that the rapid development of China's wind power market has also led to excessive competition, decline in equipment prices and creation of surplus capacity.

Table 7.13

Top 10 Chinese Wind Turbine Manufacturers (as in 2009)

Name	Newly Installed Capacity (MW)	Market Share (%)	Cumulative Installed Capacity (MW)	Market Share (%)
Sinovel	3,495	25	5,652	21
Goldwind	2,722	19	5,343	20
Dongfang	2,035	14	3,328	12
United Power	768	5	792	3
Mingyang	748	5	895	3
Vestas	608	4	2,011	7
XEMC Wind Power	454	3	–	–
GE	322	2	957	3
Suzlon	293	2	605	2
Gamesa	276	2	1,828	7
Others	2,079	15	3,814	14
Total	13,803	100	25,805	100

Source: Global Wind Energy Council, 2010.

Wind R&D

On the supply side, wind energy research has been an important focus of China's three key programmes: the national basic research programme (973 programmes), the national high-tech R&D programme (863 programmes) and the national key technology R&D programme. However, so far no research institute has been established with focus purely on applied technological wind energy research unlike in India, where the Centre for Wind Energy Technology (C-WET) was established as far back as 1998. At present, seven Chinese enterprises—Sinovel, Goldwind, Dongfang, Zhejiang Windey, Xiangtan Electric Machinery Corp (XEMC), Guodian, United Power and Guangdong Mingyang—essentially possess the design capability and key manufacturing technology for MW-scale wind turbines, and are in a position to carry out independent research, develop advanced design, software and control strategies, establish high-level laboratories and other R&D platforms and cultivate professional technical personnel.

With increasing competition at home and abroad, China's wind industry is seeking to secure home-grown intellectual property rights, especially for core technologies, such as electrical control systems, and improve its supply chain of offshore wind turbines and reduce the installation costs in the near future. China's offshore wind industry also faces several technical barriers. Most offshore wind projects around the world are located in shallow waters (0 to100 feet), so that the turbines are built directly on the sea floor. However, as turbines are installed farther from the shore in deep waters, more advanced technologies are needed and the construction costs increase exponentially due to engineering complexities.

Rise in Prices of Rare Earth Metals

Since an increasing number of wind power equipment manufacturers in China have started to adopt direct-drive permanent magnet technology, rare earth minerals, some of the core raw materials used to produce wind turbines are becoming a significant barrier to their development. The recent price surges have forced many industry players to halt production, or create a complete industry chain by expanding into the upstream sector. In response to the soaring raw material costs, Chinese wind power equipment manufacturers are beginning to adopt advanced processes to reduce their reliance on NdFeB (neodymium iron-boron magnets, a crucial component of wind turbines), which are 10 times more powerful than conventional magnets. Chinese manufacturers are also reducing their process costs while improving the performance of their turbines by spending more on R&D. Gaining direct access to upstream resources (rare earth mines) is also seen as a hedge to rising costs.

Functioning of the Innovation System

Based on the results from our questionnaire, the performance of the innovation function can be summarized as follows:

- Knowledge development is being driven by incentives such as the R&D subsidies and corporate R&D efforts, and innovation pressure, such as fierce price competition and the need to improve turbine performance.
- Knowledge diffusion has been best served by a 20-fold increase in wind power over the last four years. Clearly, the Wind Mega Bases were driving this growth and the supply side policies were channeling a lot of this demand towards domestic suppliers.
- Legitimacy has gained momentum due to the framework regulation of the Renewable Energy Law in 2006 and the targets for renewable energy. The successes in building a domestic turbine industry and the increase in the targets have further reinforced the legitimacy of the technology.
- Guidance of search was not as well served as it could be. The legitimacy enforced by policies often lacked sufficient implementation. The need to improve the grid was not sufficiently communicated in the actions, supported by the early focus of the demand regulation on installed capacity instead of the electricity fed into the grid. Moreover, the sheer success and speed in market growth, which was achieved by the big state-owned electric utilities, also had its downside. The Chinese manufacturers were not forced to look into new business models, such as turnkey projects, and the tremendous demand pushed quality concerns somewhat in the background.
- Market formation is extremely strong, as the onshore market is moving towards a mass market in the face of a 100-plus GW project pipeline. However, high potential niche markets (offshore, decentralized solutions) still have to be explored.
- Entrepreneurial activity is high, with a value chain clearly dominated by domestic players. China's wind power equipment supply chain is growing and expected to become the largest in terms of output in the near future.

- The resource supply situation has improved for some resource categories (financial resources, site resources, industrial resources), while for others there are still obstacles on the path to professional industrialization (human resources and infrastructural resources).

To sum up the experience, the main force has been the strong political commitment to renewable energy. The large projects implemented to follow the targets led to increasing demand, which fostered entrepreneurial activity. Positive feedback loops have been established with knowledge development, entrepreneurial experimentation and market formation reinforcing each other. The regulatory mechanisms have successfully contributed to the emerging of a massive wind turbine and component supply chain. Target setting and massive market formation increased the legitimacy of this technology. The large electric utilities were picking up the targets and were installing capacity very quickly. However, to maintain this in the future action needs to be taken today, by both policy and (state-owned) corporations. Based on the results of the expert questionnaire, the following challenges and barriers have been identified in China:

- Integration of large-scale wind power supply into the grid without reducing quality and functionality of the electricity system, which requires investments on the one hand, but also improved grid management with regard to fluctuating wind power supply on the other;
- Lagging behind in developing technology innovations and adapting wind turbines to national conditions through indigenous R&D;
- Lack of technical standard, testing activities and certifications;
- Lack of operational and managerial experience with wind farm management, including maintenance; and
- Overcapacities in the Chinese wind energy equipment market. This refers to the numerous companies starting to enter the equipment market, sometimes with very little experience and low-quality products aiming to win on their low-cost production.

Some of the challenges relate to quality problems, which lower the full load hours per year. Furthermore, various reports are complaining about problems in connecting all the new capacity to the grid. These problems seem to be connected to a policy which was focusing on wind capacity installation instead of wind power generation, and to an institutional structure, in which large state-owned utilities which are less subject to direct market pressure than small Independent Power Producers (IPPs) are the main investors. Furthermore, this incentive system also seems to be consistent with manufacturers putting less emphasis on testing and quality checks relative to a quick increase in output of wind turbines. With Chinese manufactures still focusing on the domestic market, there was also no need to fulfill the quality standards of international customers. Finally, even if foreign licenses are taken, it takes more time to build tacit knowledge, which is necessary to come up with a high level of quality. Thus, some of the quality problems also seem to be related to the rapid speed with which the development of the wind turbine industry was taking place.

However, the Chinese innovation system is in a process of further development. With the amendment to the Renewable Energy Law in December 2009, a major step towards a generation-based performance metric was undertaken. It remains to be seen what the effect of this change on the challenges ahead will be. Furthermore, Sinovel and Goldwind have plans to develop larger turbines. This would bring them closer towards developing cutting edge innovations, and it could also make Chinese companies more competitive on international markets and on the markets for offshore wind farms.

Comparison of Chinese and Indian Wind Turbine Industries

An analysis of both countries reveals some similarities. Both have successfully built up domestic wind turbine (WT) industries. Wind markets in both countries are now in a rapid growth phase. Similar drivers are involved, contributing to energy security, environment at pressure, socio-economic potentials, political commitment and regulatory support.

The analysis also highlights some differences:

- In China, five state-owned electric utilities dominate wind farm operations and are the major customers for WTs. In India, turnkey projects have resulted in wind farm developers being a mix of WT manufacturers; independent companies and electrical utilities are not as important a factor as in China.

- Market development in China closely follows the development plans of the Chinese Government. In India, government-provided incentive structures have resulted in a wind market driven more by private sector participation.
- In China, domestic firms (overwhelmingly state-owned) dominate the market. In India, there is a higher involvement of international players in WT production and, increasingly, in financing.
- The dominant Indian WT player (40-plus market share), Suzlon, enjoyed large early successes in world markets (with a cumulative WT supply stock in China itself of 2,150 MW as of 31 March 2012), and so, was exposed early on to the requirement of international markets with regard to quality and reliability. The development in China, in contrast, has focused entirely on the home market: so far, global entry has still to be achieved even by the leading Chinese players.
- Project performance and technical standards and certification are still far more under developed in China compared to India.
- Suzlon's WT product range goes from 0.6 MW to 6.2 MW. No Chinese WT maker comes anywhere near this. Suzlon is setting up a 1,500 MW wind farm in the Kutch region of India's western state of Gujarat on a turnkey basis in four years, which it will also operate and run as a developer. There is not a single wind farm in China as large as this or that produces electric power so cheaply. Suzlon is also more advanced in off-shore WTs than any company in China.

Solar Photovoltaics

China launched its R&D and laboratory-scale 'production' programme of solar cells and panels/modules in the late 1980s. India had done likewise in the late 1970s. In fact, our public sector enterprise, Central Electronic Ltd (CEL) was the second SPV company in the world, after the US oil company ARCO (Atlantic Richfield), to enter the SPV area. By 1982, CEL had a 1 MW/year capacity large pilot production plant based wholly on Indian technology up and running. CEL went into regular commercial production with a 3 MW/year capacity in 1984.

In contrast, the Chinese SPV companies started comparable levels of production only in the early 1990s, and that too as contract manufactures to American, Japanese and German SPV companies. But with the significant growth in the world SPV market in the early 2000s, Chinese state-owned enterprises started their own independent/non-contract manufacturing units. Their technology was a mix of what they got out of being contract manufacturers using technology transferred to them by American, Japanese and German SPV companies, and by copying, reverse engineering, buying top-of-the-line manufacturing equipment from SPV equipment manufacturers in those countries, along with some of their own (Chinese) R&D.

By 2005 there were five major SPV manufacturers in each of the two countries with a total annual production capacity of around 50 MW each. As for market orientation, all Chinese companies were totally export-oriented in regard to their solar cells and panels. In contrast, the Indian companies were much more domestic market oriented with the export–domestic ratio being around 20:80.

In regard to output/production composition also, there was a major difference between the Chinese and Indian programmes: the Chinese companies overwhelmingly produced only cells and panels, whereas the Indian companies went on from panels to the design, engineering, installation and commissioning of complete SPV power systems/sources. This character of the Indian SPV industry made it completely capable of meeting specific customer needs across a wide range of customers/applications. In the case of China, such systems activity was practically zero.

With the export orientation the Chinese had adopted in so many consumer product industries, supported by massive Chinese government subsidies and a very favourable US Dollar–Chinese Yuan exchange rate, Chinese SPV products became very cheap and gained great export advantage in the markets of North America and Europe.

On the other hand, as of 31 March 2010, India's Central Electronics Ltd (CEL), for example, had designed, engineered, produced, installed and commissioned 477,650 SPV systems of 30 different types for rural, remote area and industrial end users. This is the largest number of SPV systems produced, installed and operationalized by any single company anywhere in the world. The corresponding data for the by-now 10-company Indian SPV industry as a whole, comes to around 1.9 million SPV systems involving a total solar panel mega wattage of around 660. A particularly unique example at the world level is the ₹ 320 crore (US$ 64 million approximately) order for solar powering of lights, fans and computers in 15,500 Panchayats (village councils) and 1,100 Panchayat Samitis in Rajasthan. This single order jointly placed by the Central Ministries of New and Renewable Energy and Tribal Welfare and the Rajasthan Government, of the order of 24 MW, is the largest single SPV order placed on any company by any customer in the world. This outstanding achievement was made by the public sector company, Rajasthan Electronics and Instruments Ltd. (REIL), headquartered at Jaipur.

In contrast, the Chinese SPV manufactures have come up only in the last five or six years, and they are unable to give the 25-year warranties that the US, Japanese, European and Indian manufacturers provide. Moreover, Chinese cells and panels have several technical deficiencies, for example, low efficiencies and high cell breakages both in transit and during moduling (see Tables 7.14 and 7.15). The 'top10' Chinese cell and panel manufacturers are, nevertheless, able to sell their cells and panels in large quantities in the world market because of their very low prices. This low price factor is, in turn, due to the large capacities of the cell and panel plants that the Chinese have been able to set up (see Tables 7.16 and 7.17) and massive Chinese government subsidies.

Table 7.14

Global Top Five Solar Cell Manufacturers (in 2011)

Company	Capacity (MW)/Yr	Country
Suntech	2,400	China
JA Solar	2,100	China
Trina	1,900	China
Yingli	1,700	China
Motech Solar	1,500	Taiwan

Source: Global Solar Energy Association.

Table 7.15

Global Top 10 Solar Panel Producers (in 2011)

Company	Capacity (MW)/Yr	Country
Suntech	2,400	China
LDK	2,500	China
Canadian Solar	2,000	China
Trina	1,900	China
Yingli	1,700	China
Hanwha Solarone	1,500	China
SolarWorld	1,400	Germany
Jinko	1,100	China
Sunneeg	1,000	China
Sunpower	1,000	USA

Source: Indian Solar Energy Society.

Table 7.16

Efficiency, Quality and Price of Solar Cells Made by Different Chinese Manufacturers (over 2009–10 to 2011–12)

Manufacturer	Efficiency (in %)	Breakages & Other Defects (in %)	Avg. Price (₹/Wp)
1	14	0.8	55.00
2	13.8 to 14.9	5.0	40.00 to 52.00
3	13.8 to 14.9	0.8	75.00 to 95.00
4	13.4 to 15.0	2.9	47.00 to 48.00
5	16.5	0.7	46.00
6	13.8 to 14.3	3.5	51.00 to 58.00

Source: Indian Solar Energy Society.

It is only when non-Chinese users take up production of panels and systems respectively using the relevant inputs, that the 'customers' face the serious and severe 'pinches' which can be seen from the above tables. However, by then, it is too late for the technical performance of those intermediate and final products to be corrected in any way. Customers

Table 7.17

Efficiency, Quality and Price of Solar Panels Made by Different Manufacturers (over 2009–10 to 2011–12)

Year	Manufacturer	Efficiency (in %)	Breakages & Other Defects (in %)	Avg Price (₹/Wp)
2011–12	Chinese Sources	13.8 to 16.5	3.44	40.00 to 50.00
	Other Imported Sources	15.5 to 16.6	0.48	40.00 to 50.00
2010–11	Chinese Sources	13.4 to 13.8	7.70	48.00 to 58.00
	Other Imported Sources	14.2 to 15.9	0.65	50.00 to 60.00
	Indigenous Sources	14.5 to 15.0	0.60	65.00 to 70.00
2009–10	Chinese Sources	14.0 to 15.0	1.10	45.00 to 75.00
	Other Imported Sources	14.5 to 16.0	1.10	55.00 to 80.00
	Indigenous Sources	14.0 to 14.5	0.30	45.00 to 50.00

Source: Indian Solar Energy Society.

Notes: (i) Chinese Sources are: M/S Ganri Energy, M/S Ningbo, M/S Yat-Sen, M/S NBS Energy, etc.
(ii) Other Imported Sources are: M/S Microsol Intl, UAE; M/S Q-Cell, Germany; M/S Schott Solar, Germany.
(iii) Indigenous Sources are: BHEL, Moser Baer, India.

therefore, have no option but to just throw the relevant systems and panels away!!

Moreover, CEL, India's top SPV company, has successfully transferred its superior SPV Systems and Panels technology to Syria, Sudan, Mozambique and Cuba and put up turnkey plants there, which China has not succeeded in doing. The Mozambique contract, secured as recently as March 2012, is for a plant to make systems and panels of 5 MW/year capacity. It was secured in the teeth of intense international competition for a contract value of US$ 15 million, one of the largest in the world.

Table 7.18 indicates the shipments (in MWs) from China in the years 2009 and 2010. However, the limping US economy followed by the massive Euro Zone economic crisis of the last 18 months has taken the bottom out of the massive Chinese exports of those two years. According to a November 2011 World Bank report, more than 50 per cent of Chinese

SPV companies have had to go so far as to actually closedown their plants. Production in just over six months has dropped from 25,000 MW/year to 15,000 MWs i.e., by around 35 per cent. This has forced China to look inwards at their domestic SPV market. The Chinese government announced in December 2011 that it would now target 20,000 MW of SPV-based grid-connected power plants at home by 2020.

Table 7.18

Shipments (in MWs) from China in the Years 2009 and 2010

Company	2009 Shipment (MW)	2010 Shipment (MW)
Suntech (China)	704	1572
JA Solar (China)	520	1464
First Solar (USA)	1100	1411
Yingli Solar (China)	525	1062
Trina Solar (China)	399	1057
Motech Solar (China)	360	924
Q-Cell (Germany)	586	907
Gintech (China)	368	827
Sharp (Japan)	595	774
Canadian Solar (China)	193	588

Source: Solar Energy Association of China, 2011.

India has been wiser in having, from the beginning, a domestic market-oriented SPV policy launched as far back as November 2009. The Jawaharlal Nehru National Solar Mission targets not only 20,000 MW of SPV-based grid-connected power plants by 2020, but also, additionally, 5,000 MW off-grid, rural-oriented decentralized SPV systems. Moreover, in the first two years of the mission itself, 1,000 MW of grid-connected and 500 MW of decentralized systems were operationalized.

A spectacular achievement of the Indian Solar Mission has been that by using market forces in the form of two rounds of competitive bidding to determine the tariff for utility scale grid-connected SPV power plants, the tariff has been reduced from an initial ₹ 17.5/unit to ₹ 8.6/unit of power. China is struggling to match India's ₹ 8.8/KWh tariff and has, therefore fallen back on a Government-fixed, arbitrary tariff of ₹ 7.5/unit. In a market economy like India's, such an unrealistically low tariff would make the whole SPV industry go bankrupt. But in China the government can fix any price!

Telecom Industry in India and China: A Comparison†

Telecommunication Industry in India

India was the first developing country to go in for the manufacture of telephone switching equipment (exchanges) when the government set up, as early as 1948, the state-owned company Indian Telephone Industries (ITI) at Bangalore to manufacture electro-mechanical switches and radio transmission equipment to interlink the exchanges. This programme was launched on the basis of technology licenced by ITI from the then well-known UK company, Standard Telephones and Cables (STCs).

These internationally techno-commercially competitive exchanges at the time used the so-called 'step-by-step' electro-mechanical Strowger technology. The ITI provided, through its then customer, the Department of Posts and Telegraph, a prompt and good quality service to its customers and also rapidly indigenized the exchanges, thereby sharply reducing the per line cost by 15 to 20 per cent.

Then in 1962–63, the ITI went into a second technology licence, this time with the US-based trans-national corporation, International Telephone and Telegraph (ITT) for the Crossbar technology-based exchanges, supposedly for better performance and reliability than the Strowger technology.

However, when introduced in the Indian telecom network, the ITT's Crossbar Exchanges gave numerous problems right from the word go. Some of the problems were due to the design and engineering of ITT's Crossbar exchanges; others were due to an incompatibility between the

---

† The author acknowledges with thanks the use by him, in some parts of this section, of the passages which had originally appeared in Wikipedia at http://en.widipedia.org/ wiki/Telecommunications in the Peoples Republic of China # cite note-o.

capabilities of Crossbar technology and the then existing character of our telephone network, subscriber behaviour and the fact that there were not enough numbers of Crossbar Exchanges in the DOT network to meet the colossal customer demand of the 1960s and 1970s. It took till the end of the 1970s for the crossbar exchanges to give trouble-free service.

Recognizing that all types of electro-mechanical exchanges were rapidly being superseded the world over by electronic ones, the Telecom Research Centre (TRC) of the DOT-started R&D on Stored Programme Controlled (SPC) analogue electronic exchanges as early as 1973–74. By the end of the 1970's, the TRC's 250-line prototype exchange had been designed, developed, fully engineered and extensively field-tested and proven in the DOT Network and was ready for: (a) scaling up in capacity (number of lines and call handling capacity), and (b) bulk commercial production by ITI.

Meanwhile, all the stakeholders in the matter of the choice of technology for the next generation of electronic exchanges—the DOT, the TRC itself, the Department of Electronics (DOEs) and the Planning Commission—came to the unanimous decision that India should skip the step of analogue electronic exchanges taken by the industrially advanced countries and proceed directly to the ultimate digital technology. The DOT and the DOE therefore jointly set up a high level committee chaired by the former Defence Secretary, Harish Sarin, to draw up a roadmap for that purpose. A former Defence Secretary was chosen because the Army had already got developed and produced all-digital exchanges by another state-owned company, Bharat Electronics Ltd (BEL) and inducted into its Radio Engineering Network as far back as 1980–82.

While the Sarin Committee was at work, an NRI, Satyen Pitroda, who owned and was running a digital electronics communication company in Chicago, wrote to Prime Minister Indira Gandhi proposing that he was ready to relocate himself along with several of his engineer colleague to India to form the nucleus of a new autonomous organization, to be set up and funded wholly by the Government, to design and develop a whole family of state-of-the-art digital electronic exchanges fully optimized for Indian conditions for the DOT network. Pitroda emphasized in his letter: (a) the proven track record of himself and his team in such a task; and (b) that the total cost of the 'Mission' he was offering to undertake would be a fraction of what ITI would have to otherwise pay to acquire this technology from any of the European multinationals. Pitroda concluded his letter by offering to come to Delhi to make a detailed presentation of his proposal to the PM, the Communications Minister and anyone else she may wish to invite.

Mrs Indira Gandhi was greatly enthused by Pitroda's offer and replied to him herself indicating a date and time on which he could make his presentation. In the event, she invited the Ministers of Communications, Defence, Planning and Finance (she herself was the Cabinet Minister for Electronics), the secretaries of those ministries, members of the Telecom Board and the Director General TRC, Member (S&T) in the Planning Commission and her scientific adviser, Professor M.G.K Menon, Harish Sarin and the key members of his committee and a few other policy makers and technical experts.

Pitroda's proposal was accepted in toto, and an autonomous institution, called the Centre for the Development of Telematics (C-DOT), was set up by the Government in August 1983. By end-1985 i.e., in record time of merely 18 months, the first exchange—the 128 line Rural Automatic exchange—was developed and field-proved. With that achievement under its belt, C-DOT never looked back. Over the next decade, the Centre designed and developed a whole family of exchanges based on its state-of-the-art technology ranging in capacities from 250-line RAXs to 40,000-line MAX exchanges for the Metros. Over 1995–2006, the 10 technology-licensee companies of C-DOT had produced, and DOT had inducted into the telecom network, around 42 million lines of exchanges constituting about 50 per cent of the total number of lines in the network. The money value of those exchanges was ₹ 72,000 crores (approximately US$ 14 billion)! All of them worked beautifully carrying not only voice but also data and video. They were also fully Integrated Services Digital Network (ISDN)-capable, like the exchanges made by the transnational corporations (TNCs). Moreover, the cost of production per line of the C-DOT exchanges of all sizes was one-third less than those of the TNCs. The total expenditure on R&D, design, development and engineering and also a model/pilot manufacturing plant of ITI in Bangalore—to make the exchanges in small numbers for operational trials in the DOT network—was a mere ₹ 1,000 crores over 12 years!! This expenditure may be compared with the US$ 700–800 million (about ₹ 4,500 crores) spent by the TNCs and their governments (on subsidies, etc.). What is more, the 7 C-Dot licensee companies also exported the exchanges made by them to around 22 countries, from Russia to Vietnam, Malaysia and the Philippines, to all the North African countries, to Nigeria and many countries in West Asia, to Venezuela, Cuba, Colombia and Paraguay. Furthermore, C-DOT set up manufacturing plants in Nigeria, Vietnam and Tunisia based on the export of the C-DOT technology.

The Chinese never come close to matching C-DOT's magnificent and unprecedented work in our entire history of hi-tech R&D, mass manufacturing and some other major Indian technological and industrial achievements in Telecom. Between 1992 and 1996, a band of talented communication engineers at IIT Chennai, led by an outstanding engineer, Professor Ashok Jhunjhunwala, designed, developed, engineered and field-proved a highly innovative telecom technology called Wireless In Local Loop or WILL. This technology actually foretold the switching servers and networks of cellular phones. It replaced cumbersome copper and optic-fibre cables to interconnect exchanges by high reliability, lower cost and quick to set up wireless links.

Fortunately, not only the Department of Electronics but also, ISRO, CSIR, and even a somewhat reluctant DOT, rallied around Jhunjhunwala and his team and the WILL technology was transferred to the same seven large lead manufacturers making C-DOT exchanges. After much battling, the DOT was forced to accept WILL and by 1998–2000 large numbers of WILL systems, totaling some ₹ 30,000 crores in value, became operational in the DOT network. What is more, IIT Chennai strongly supported engineer-member of Jhunjhunwala's team to set up a number of 'spin-off companies' making critical hardware and software sub-systems for complete WILL systems. Many of the WILL technology licensee companies also exported WILL systems to a number of the technically advanced developing countries, such as Malaysia, Indonesia, Algeria, Morocco, Tunisia and Egypt. Indian companies also transferred WILL technology to major state-owned telecom manufacturing companies in Algeria and Tunisia, and those 'second generation' licensee companies not only progressively started meeting their own domestic need of WILL, but also exported their WILL telecom system to other countries in north and west Africa.

However, then in 2005–06–07 came the revolutionary and 'discontinuous', not 'evolutionary' new technology of cell phones. This emergence, and the very rapid growth of cell phones, initially dealt a body blow not only to C-DOT's land-line exchanges but even the wireless links-based WILL. But C-DOT has quickly recovered and developed a whole range of new telecom technologies. Some of the major ones are:

1. **GPON:** C-DOT's Gigabit Passive Optical Network (GPON) is a future-proof, end-to-end solution for rural and urban broadband connectivity. Fully compliant with ITU-T G-984 standards, it can be used to launch FTTH (Fibre-to-the-Home) services, bringing triple play (voice, video and data) to the subscribers through one single optical fibre pipe.

2. **NGN:** The building blocks of the core of Next Generation Network (NGN) viz., Soft Switch, Media Gateway, Signaling Gateway, Session Border Controller, Router, etc. for providing Carrier-Grade service, have been developed in C-DOT, and the technology transferred to Industry.

   Simultaneously, a solution (MAX-NG) for converting all TDM switches (MAX and RAX exchanges) based on C-DOT technology to IP-enabled exchanges where the calls will be switched by the Soft Switch, has also been developed by C-DOT and has recently fully passed DOT field trials. It will catapult existing wire-line infrastructure seamlessly to an IP network while ensuring investment protection and greater customer satisfaction. Again the technology is being transferred to Industry by C-DOT.

3. **BBWT:** Broadband Wireless Terminal (BBWT) is a compact wireless device developed in C-DOT to take broadband to rural and remote areas. Based on WIFI technology, it operates in unlicenced band and makes use of innovative cognitive radio techniques to select the best channel in a potentially noisy environment.

4. **MCA (MISSED CALL ALERT):** C-DOTs Missed Call Alert (MCA) system is being used widely by BSNL in their GSM network. A very cost-effective, yet reliable and feature-rich system, C-DOT MCA's hardware and software were developed entirely in-house at C-DOTs. Some 22 systems are in operation with BSNL all over the country, with the first system having logged over five years of fault-free service in the network.

5. **SG-RAN:** Shared GSM Radio Access Network (SG-RAN) is an innovative product from C-DOT which permits sharing of active

radio infrastructure by up to three different operators, each using its own spectrum and service features. It has been developed to reduce COPEX and OPEX so as to expand mobile telephony to rural, remote and essentially low-ARPU areas.

Totally Unnecessary Telecom Imports

However, due to massively wrong GOI policies, the situation has got so bad that our indigenous telecom hardware industry which, pre-1991, met 70 per cent of national needs and enabled exports on a reasonable scale now meets only 20 per cent. This data is from a mid-2011 TRAI Consultation Paper which also recommended that the Government should immediately announce that its policy in this area would now be that DOT would not allow any telecom equipment with less than 60 per cent Indian content by 2015 and 80 per cent by 2020 to be part of our telecom network.

Soon after the TRAI paper came out, Satyen Pitroda, who was now Adviser to the Prime Minister on Information and Communication called a major press conference on 18 July 2011 where he made public, for the first time, that in 2008–09 imports of all types of electronic hardware had cost ₹ 46,200 crores! He urged the government to launch, on top priority, a crash programme for drastically and rapidly reducing such huge import levels through a number of measures all of which involved end-product import substitution and quality-enhance levels of local content. He gave the example of the ordinary cell phone which had only 25–30 per cent, local content, whereas infinitely more complex military wireless sets made by BEL had 90 per cent Indian content. He warned the government that if the necessary measures were not adopted, by 2025 India's electronic equipment and component imports would exceed the value of our present oil imports of ₹ 170,000 crores.

Surprisingly, the so-called cell phone handset "manufacturers" in India have become aware of these dangerous-looking effects of the atrociously low levels of indigenization only now, though they started assembling handsets many years ago. Nokia India, for instance, started such assembly as far back as 2006; but even now, six years later their indigenization level is only 25–30 per cent! And that 25–30 per cent consists of low-technology simple items like back covers, key mats (not keys themselves) and battery chargers! Not even the main printed circuit board (PCB) or the window display, or the many passive electronic components populating the PCB! However, the assessment of some companies is different. They say indigenization, apart from saving costs and thereby increasing profits, also ensures flexibility in responding to the changing market. For example, the MiHal vice president for manufacturing says: "Currently we keep inventory of two months. This will go down to one week if we indigenize. This means a cost saving of 5–10%, which is pretty large in this business."

Telecommunication Equipment Industry in China

As far back as 2002, China became the world's largest telecom market. As of 2009, there were 380 million fixed landline subscribers and 750 million mobile subscribers. By 2011, the number of landline subscribers had not changed much, but the number of mobile subscribers had jumped to 930 million. The corresponding figures for India in 2011 were 92 million fixed/landlines and 875 million mobile subscribers.

Until about 2008–09, China was the fastest growing cell phone market in the world. However, in 2010 and 2011 India had taken its place with about 15 million cell phones being added every month or around 180 million phones per year. The growth rate in India had become 28 per cent compared to China's 14 per cent! However, in both countries cell phones have yet to penetrate the poorest and the most remote areas. For instance, in September 2011, as stated by India's Minister of State for Communications in a reply to a Parliament Question, 37,184 villages out of a total number of 593,000 were yet to be provided with mobile telephone service.

Although the DOT had created, many years ago, a huge Universal (Telecom) Obligation Fund for mobile telephony service providers to have this service in rural areas, not a single private sector service provider has fulfilled that obligation even to the extent of a small percentage of the need, even though the said service was to be financed by government. And government, true to form, instead of issuing showcause notices to the 20-odd private service providers why their licences should not be cancelled, fell back on the ever present and ever 'orderable' public sector mobile (and fixed) service provider BSNL to take on the legal obligation of the defaulting private sector, of course, with financing from the Universal Access Fund! The much maligned BSNL did a splendid job of covering in a mere three years all 650 district headquarters (which the private sector was not willing to touch!) apart from 33,620 metros, Type I and Type II cities and towns. In China, where all telcos are state-owned, it is not surprising that the rural coverage is 96 per cent compared to our 94 per cent—indeed it is surprising that the Chinese coverage is not 100 per cent!!

In China, the structure of telecom service provision is exclusively Chinese and by wholly state-owned companies: two fixed line operators, with nationwide licences—China Telecom and China Netcom; two mobile carriers—China Mobile (using GSM technology) and China Unicom (using both GSM and CDMA technologies) and a China Satcom and a China Railcom.

Foreign participation in the provision of telecom services was totally banned till 2009. However, in the last two years, that policy has been partially changed to permit Chinese state and foreign carrier joint ventures but with a mandatory condition that the Chinese state partner would have majority equity holding! Despite this condition, the US and all the major European countries, as also Japan and South Korea, did manage to form such joint ventures. Such was the commercial pull of the world's largest telecom market! But within the last two years, most of the joint ventures had ended in 'divorce' with disastrous losses for the foreign companies.

The story is much the same in China's telecom equipment manufacturing sub-sector. With strong policy, financial, preferential procurement and many other explicit modes, there are many implicit elements and areas of government support to national and very substantially state-owned companies which dominate the telecom equipment production. While notionally they compete with foreign equipment suppliers, there are only a very few foreign equipment manufacturers in the Chinese market: the whole deck is massively in favour of the Chinese state-owned companies. At the last count, in early 2010 there were around 20 'Chinese' (all state-owned) telecom equipment manufacturers. It is estimated by several telecom industry consulting firms—both within and outside China—that these 20 constitute and control 95–98 per cent of the Chinese telecom equipment market. The lead companies like Huwaei, ZTE and Datang are very large with annual revenues in the range of US$ 80–100 billion. Moreover, all three have, in turn, several wholly-owned subsidiaries of theirs in around 30 countries. For example, in India they have been cornering around 50 per cent of the 2008–10 telecom equipment imports worth ₹ 46,000 crores!

But where do these three lead companies and the rest of the 20 companies acquire their equipment designs and manufacturing technology? The usual Chinese mix of technology sources across economic sectors are: some genuine R&D, copying, reverse engineering and outright stealing in both embodied and disembodied forms of technology. Backed massively by the Chinese state, the companies totally ignore WTO provisions!

Telecom hardware and software producers, even if allegedly meant or claimed to meet civilian, commercial needs of telecom service providers, have, by their 'hidden' manning structure and intrinsic technical characteristics, been posing India serious security-related concerns. For example, all the top three Chinese companies—Huwaei, ZTE and Datang—were founded and are headed by senior generals of China's PLA, and most of their technical work force consists of military communication engineers—in R&D production and marketing.

China, like all other countries, also used fixed/landline switching and transmission equipment to build up its telecom services. The production of landline switching equipment started as far back as 1960, based on technology from the former Soviet Union. The switches went from electromechanical to electronic, and from analogue electronic to digital by 2006. Similarly, the transmission equipment went from copper coaxial cable to optic fibre cable and the radio relay from UHF and microwave analogue to digital equivalents, and then to satellite by around 2005. The technology for all those products was developed in China by copying and 'reverse engineering'. Production of this entire product range was undertaken by six very large state-owned companies over 1960–2006. At the end of August 2006, China had approximately 370 million lines of such equipment in its network.

Then, in 2003, the Mobile Radio Equipment came in and the Chinese Government, the service providers and customers alike grasped it with both hands. The government initially reduced customs duties on finished mobile phones to 5 per cent so as to flood the market with imported mobile phones mostly from Europe. But, in barely three years i.e., by early 2007, there were 440 million mobile subscribers with the number growing by 1.25 million new subscribers every week by 2008.

In 2004, the Chinese government adopted a number of major policy measures to build up a domestic mobile phone manufacturing industry. Three major, fully vertically integrated companies, covering both network equipment and handsets, were established, all promoted and staffed from top to bottom by engineers from the Peoples Liberation Army (PLA) of China. These were: Huawei, ZTE and Datang. The three were soon followed by six more—Shanghai Bell, Amoi, Konka, Ningbo, Bird and Kejan. Concurrently almost all the leading international suppliers of mobile handset—Nokia-Siemens, Motorola, Ericsson and Samsung, and equivalent network equipment suppliers—Alcatel-Lucent, Cisco, Ericsson Nortel and Siemens—all operate in the Chinese market.

Huawei and ZTE were founded way back in 1980. Since their inception they were nurtured by China's national policy to build its networks only through state-owned companies. Both companies were handed out huge contracts from China Mobile to cover the whole country with wireless service. From that monopoly base, the companies expanded into the emerging markets in Africa and Latin America around the year 2000, and then into the developed markets. These equipment manufacturing companies also get help in the form of multi-billion dollar loans from the Chinese state banks; in turn, the companies offer their operators incentives to buy their equipment. Such loans have been attacked by their foreign rivals, but to no avail.

It is estimated that in 2006, the total size of the Chinese telecom equipment market was worth around US$ 30 billion. However, the levels of import and export were significantly different. While exports were US$ 66 billion import were a mere US$ 12 billion. Moreover, while the rate of growth of the overall market was 15 per cent to 20 per cent, the imports growth rate was only 14 per cent over 2005–09. Moreover, the European companies mentioned above dominated imports, with the US, Japan and South Korea having only around a 6 per cent market share.

Most importantly, by 2009 the Chinese manufacturers had come to dominate the market with a 72 per cent share of the Chinese network equipment market, thanks to low-cost production and a good level of technology and rapidly increasing local content. By 2011, Huawei had become the world's number three maker-and-seller of network equipment, and ZTE had become number five. They have also made deep inroads into Europe and the rest of Asia, except for Japan and South Korea. The US, European and South Korean companies—Nokia-Siemens, Ericsson, Motorola and Samsung—however, continue to have a vice-like grip of the Chinese mobile handset market as they do of the global handset market as a whole.

Starting 2010, Huawei and ZTE started attacking the world's largest telecom market, the US, where they met a major non-commercial impediment. The US Congress, the Pentagon and the CIA, raised concerns about having these Chinese companies in the US telecom networks, because both the Chinese companies were owned and operated by the PLA. A major US telecom company, Sprint Nextel explicitly excluded Huawei and ZTE from a recent US$ 5 billion tender for Sprint's network modernization. Huawei also had to give up a bid for acquisition of a US company, 3 Com, in 2008, while its recent acquisition of US technology firm 3 Leaf is currently under White House review, both due to security concerns having been raised. So, all in all the Chinese companies are facing a brick wall in the US. Indeed, we in India also have allowed the Chinese companies to operate here, but only under very stringent security-related conditionalities.

Both Huawei and ZTE are now very large companies with sales volumes per year of around US$ 100 billion each. Their US and European rivals complain that they are not competing with the two Chinese companies on a level playing field for all the reasons indicated above. Indeed, they go further and say Huawei and ZTE are not normal companies: they are 'China Inc' and part of a long-term strategy by the Chinese Government to acquire absolute dominance in one market after another and above all in the total telecom technology space! These companies play by a made-for-themselves set of rules, on everything from intellectual property to labour standards and to corporate reporting to their Master—the Chinese State.

What Must India Do To Learn from Chinese Strategy in Telecom? What and Why Have the Chinese Done What They Have Done?

China has used political will and market size to force foreign equipment suppliers to produce locally, thereby creating a production-supporting eco system and leveraging its strength by taking the following measures.

- The Chinese Government started, way back in the early 1990s, programmes like 'China 863' to undertake a state-directed hi-tech plan to develop, test and prove products in hi-tech economic sectors like telecom.
- Foreign mobile handset suppliers, like Nokia-Siemens (NSN) and Ericsson, were forced to manufacture in depth locally with 95 per cent Chinese staff and content.
- Massive financial support was given to Chinese telecom companies to undertake high levels of RDDE by providing public funding in the form of grants and cheap loans for production, marketing and after-sales service.
- Huge amounts of money were provided by the Government to both commercial telecom companies and government laboratories and universities for taking out international patents and building up patent data-banks. Consequently, the Chinese industry and government were able to jointly define their own 3G standards (TD-SCDMA). The Government also consciously delayed the rollout of their 3G until the requisite domestic manufacturing competences were fully developed domestically.

What are the Barriers to Indian Telecom Equipment Manufacturers and Government Emulating the above Chinese Strategy and Tactics

- The inability of large equipment manufacturers to give credit to operators.

- Currently, most equipment manufacturers offer equipment on end-to-end services models. Given the relatively small scale of their operations, the credibility and the project management skills of the small players remain a major issue for an operator.

Given that the keys to success in this industry are talent and software skills (which are abundant in India), some of the merits in producing locally are:

- The market potential for the industry in the next five years is of the order of US$ 100 billion.
- If we produce locally, we can create at least 75,000 skilled jobs and 25,000 semi-skilled jobs in five years.
- Local production will reduce the cost of operation drastically (also, the equipment can be customized as per our needs) for telecom operators, which would help to reduce the rural–urban divide.
- Given that India has to embark on its journey in creating a total telecom manufacturing ecosystem, we can control radiation and energy consumed by our products.
- Indigenous development will promote innovation and development of innovative products.

The Action Steps Which the Government of India and the Indian Telecom Equipment Manufacturers Need to Take to Overcome Problems and the Barriers Mentioned Earlier

- Create a dedicated 'R&D Fund' from auction of spectrum income and the balance from the underutilized USO fund. The fund should be chaired by some reputed industry person on the lines of UID and should include people from academia (like IIT's, IIM, etc.) and senior members of DOT, DIT, TRAI, etc.
- Like the diamond industry is allowed to import raw stones for polishing and cutting, imports of all base components (PCB and chips) and machinery should be allowed duty-free subject to reasonable value addition norms.

- Wireless equipment developed in India should be supported by allocating special research spectrum free-of-cost in a predetermined geography. These are called Test Labs.
- Provide access to Test Labs to various government agencies, like C-DOT and others and tie up with other private labs inside or outside the country to provide low cost access to testing facilities.
- Create a certification and testing laboratory on the lines of the China Certification and Accreditation Administration (CNCA). All operators should be forced to use equipment only with the necessary certification as this would help the regulatory body monitor radiation and energy consumption. The same facility can also be used to test mobile telecom apparatus which require stringent security processes (like m-payment portals).
- Encourage operators to use optical fibres for connecting towers; this will help them reduce the need of long Steel Towers (because RF waves now being used require line of sight and increased use of devices causing radiations).
- Compel MNCs to produce locally; if required, create majority Chinese-owned joint ventures with the foreign companies to produce locally.
- Incomes should not be subjected to MAT.
- CST and VAT on sale of equipment in local systems/products should be kept to the minimum (around 2 per cent).
- Encourage use of local products by forcing BSNL/MTNL (government-owned operators) to source 25 per cent of their equipment needs from local manufacturers only.
- Create collaboration between various academic institutes and government R&D laboratories to promote standardization and creating intellectual property right globally. This would also require our R&D laboratories access to the membership of various standard setting bodies globally.

Pharmaceuticals

The modern Chinese pharma industry producing Western medicines started around 1960. Over the last 50 years it has grown (on a small base) at a CAGR of around 60 per cent. However, even in 2010 the output of the industry was dominated to the extent of 70 per cent by small-scale companies, the top 10 per cent in size contributing only 20 per cent of the output, while the remaining 10 per cent of output is by Western companies.

All the 90 per cent of Chinese companies—even the top 10—are characterized by having duplicate production processes, outdated manufacturing technology and management structure. The industry also has low market concentration and weak international competitiveness, combined with lack of internationally patentable indigenously developed pharma products.

China's joining the WTO in 2001 put intense and varied pressure on its pharma industry in particular. The industry had to contend with drastically reduced protective customs-duties and free entry of Western TNCs into the Chinese pharma market. The requirement of the TRIPS regime of WTO also banned process patents making it desperately necessary for the industry to do basic and applied research on its own to discover new chemical entities (NCEs). But China totally lacked pharma R&D, unlike India which has a sound and substantial R&D base, not only in the companies but in major government labs like the Central Drug Research Institute, Lucknow, the Indian Institute of Chemical Technology, Hyderabad, the Regional Research Laboratory, Jammu, the Institute of Microbial Technology, Chandigarh and the Institute of Genomic Biotechnology, New Delhi, the Council of Scientific and Industrial Research (CSIR) headquartered in New Delhi and the National Institute of Immunology (NIL), New Delhi under the Department of Biotechnology (DBT), also in Delhi. The institutional capabilities for even minimal technological innovation capacity and capabilities, and investment in R&D and new product development, were major handicaps for China. What is more, China has a few major pharma companies, of which not even a minimal number had achieved the necessary economies of scale in production, marketing and human and financial resources. Most Chinese drug producers had got used to relying on repetitive production of low value-added bulk drugs and imitation drugs in a highly protected domestic market with zero exports.

Structure of the Chinese Pharma Industry‡

The first problem is the question of company size. Even the top-20 selling companies (out of the total 3,500 companies in 2010) have sales barely exceeding US$ 100 million (₹ 500 crores) per year.

Table 7.19

Drugs Developed by Indian Companies through Local R&D

S. No.	Name of Company	Products/Drugs Developed/ Being Developed in 2010	Remarks
1	C&D Pharma Technology Holdings Ltd. (SIN: COPT)	Developing a range of drugs addressing acute medical afflictions, for example, anti-cancer, anti-TB, anti-ageing, anti-obesity. One Hepatitis-B drug abroad spectrum anti-bacterial antibiotic.	Eighty in-course researchers specialized in chemical synthesis, drug formulations and preparation methods, clinical research, product registration, development of manufacturing techniques and pharma analysis. One of the few qualified Chinese pharma companies with expertise to provide contract research services to the growing number of TNCs seeking to outsource their R&D activities to China. Also provides such contract services abroad/outside China.
2	Shijia-Zhuang Pharma Group (Northeast China)	One of the largest pharma industries in China. Very typical of post-2005 drug industries in China in late 2005. Group announced official launch of its investigative drug Nutro-Butyl- Phthalide (NBP). It obtained domestic patents for the drug from Chinese	The group uses three routes to develop new drugs. First, develop them in collaboration with universities and research institutes. Second, apply for generic drug right before the (domestic) patents of patented drugs expire; and thirdly, modernize traditional Chinese medicines (TCM) i.e., develop

‡   The information and data in this section in regard to China is based on various publications of the Ministry of Chemicals and Pharmaceuticals of the Chinese Government; and likewise in the case of India in addition to those of the Indian Drug Manufacturers Association and the CSIR and the DST of the Government of India.

S. No.	Name of Company	Products/Drugs Developed/ Being Developed in 2010	Remarks
		Academy of Science for US$ 4 million and then spent US$ 6.3 million on clinical trials. However, even as of 2010 the drug was not in production nor available in the Chinese markets.	TCMs in the same quantitative way as used in developing modern synthetic drug-based ways. The third way was used in the development of NBP, a TCM extracted from celery seeds and effective as an anti-bacterial and anti-fungal drug/medicine.
3	Wuxi Pharmatech (Cayman Islands Inc.)	A Chinese–US JV operates, through its Chinese subsidizing as a pharma and biotech R&D outsourcing company in the PRC. It provides a portfolio of labs manufacturing services in the drug discovery and development.	

S. No.	Name of Company	Drugs Produced in 2010	Products /Drugs Developed/being Developed in 2010	Remarks
4	Harbin Pharma Group Co.	Synthetic Chemistry-based generics for some 20 of the most common diseases in China No product which is special or new.	The US company capital infusion will allow Harbin to expand its R&D efforts. Current R&D spending is 5 per cent of sales, exceptional for a Chinese drug maker.	A US$ 250 million capital infusion from two foreign investors planned by end 2012. However, the 5 per cent sales on R&D is only one third of what most Western JNCs spend, Harbin is slowly moving however, from predator to prey.

(continued)

Table 7.19

(continued)

S. No.	Name of Company	Drugs Produced in 2010	Products /Drugs Developed/being Developed in 2010	Remarks
5	Sinovac Biotech Ltd.		A significant development is that the company has launched marketing BILIVE in 2010. The product is a combined hepatitis treatment. Sales expected in 2012 in terms of time are a mere US$ 6.5 million/year at GSK. GSKs BT equipment called Twinrix	A&B vaccine developed by Chinese scientists and has only one directly competing vaccine by GlaxoSmithKline (GSK), called Twinrix which is not available in China. So, Sinovac has a monopoly; moreover Twinrix sells at a much higher price than that of BILIVE in countries where it is sold. Further, Sinovac is a world leader in SARS vaccine
6	Zensur S&T Co. Ltd.	A biotech pharma co. well versed in the demands of the global market with high-tech innovation base- high profit potential	Devoted to R&D of new drugs with its own drugs with its own IP Long focusing on the R&D of anti-tumour drugs and anti-heart failure drugs. Based on Innovative Theory, the company has developed two drugs—Recombinant Human Erb 83 fragment injection, a therapeutic vaccine against tumors. Both have undergone clinical trials in 2006.	

Despite the WTO's TRIPS conditionalities and rules, even today Chinese production is dominated by its non-branded generic industry using basic technology and simple heavily-manual production methods. Domestic pharmaceuticals are technologically far behind not only the Western countries and Japan but also India. Forty per cent of the total productions of pharma-producing companies are not just state-owned by the Central government but even at lower levels by the provincial governments. They are characterized by very poor market linkages and hence by over-production and heavy losses on a continuous and recurring basis.

Counterfeit drugs and production are a very serious problem in the Chinese pharma sector. For example, 80 per cent of the drugs consumed in China's rural areas are both counterfeit and of sub-standard quality.

Apart from the 36 per cent state-owned companies, 35 per cent are what the Chinese call 'privately'-owned. But their definition of 'private' is radically different from what the rest of the world understands by the term. They are really public sector companies, except that the shares of those companies are owned by the managers, engineers and workers— all of whom are government employees. So, it is only the remaining 29 per cent companies in the Chinese pharma industry, which are genuinely private; but then all of them are western TNCs.

As for the character of the product-mix of the Chinese pharma industry, 65 per cent is synthetic chemistry-based, 21 per cent is traditional Chinese-medicine-based and the remaining 14 per cent is biotech-based.

Domestic Chinese Companies Doing R&D

Of the top-20 Chinese companies with 2010 sales of between US$ 100 million (₹ 500 crores) and US$ 500 million (₹ 2500 crores), only seven companies do any genuine/realistic R&D. The names of the companies and the drugs they have developed through local R&D are indicated in the Table 7.19.

As for foreign companies doing R&D in China, a poll of 33 such companies in 2009 revealed that only 7 out of the 33 had R&D centres in China. However, collection of more detailed information on the seven has revealed that what the companies called 'R&D' was actually dominated by clinical trials. The remaining 26 of the companies polled stated that they were not doing R&D of any kind in China. Of the seven, the maximum number of researchers in any one of them was only 40!

The Indian Pharma Scene

The origins of the Indian pharmaceutical industry go back to the 1920s. Over five years in that decade, Professor Prafulla Chandra Ray, a distinguished chemist at the Calcutta University set up, with the political and financial support of the industrialist Sir Ashutosh Mukherjee (who also supported C.V. Raman and Jagdish Chandra Bose) three pharma and chemical companies—Bengal Chemicals, Bengal Immunity and Bengal Biologicals—to manufacture with the knowhow he had developed at the university. These companies produced a range of chemicals and drugs which are marketed through hospitals. It was a brilliant pioneering achievement. Then, in the 1930s and 1940s, Glaxo and Mey and Baker, two UK firms, came to the country and started manufacturing several over-the-counter (OTC) drugs and sulpha drugs.

With the attainment of Independence, a medical doctor from the Army Medical Corps, Major General S.S. Sokhey and the WHO set up the nation's first antibiotics production company at Pune, named Hindustan Antibiotics Ltd (HAL). Towards the end of the 1950s and in the early 1960s another public sector company, Indian Drugs and Pharmaceuticals Ltd (IDPL), was set up with technology and technical assistance from the former Soviet Union. This company had three plants: an antibiotics plant at Rishikesh, a synthetic drugs plant at Hyderabad and a medical instruments manufacturing plant at Chennai. It was HAL and IDPL which formed the crucible from which came the founders of many of the leading private sector pharma companies of today, for example, Dr Angi Reddy, who set up India's second largest pharma company (Dr Reddy's Laboratories at Hyderabad), had worked for 10 years in IDPL'S synthetic drugs plant established in the same city.

Today, our pharma industry consists of 250 large companies and plants, that control 70 per cent of the market and production (with the top 10 having around 10 per cent of the market) and about 8,000 small and medium scale units contributing the remaining 30 per cent. Technologically strong and totally self reliant, the Indian pharma industry has very low costs of production and R&D and innovative scientific manpower. It maintains close connections with the relevant government laboratories, for example, of CSIR, and with a wide range of universities. It has a large positive trade balance i.e., very large exports, despite meeting 95 per cent of the nation's needs of medicines. Of the 250 large units, 200 have US Federal Drug Administration's approval of their manufacturing facilities, which is more than any other country outside the US. The top-20 companies are active in generics exports to 130 countries with a global market share which has climbed steeply from 4 per cent in 2004 to 33 per cent in 2007 and 47 per cent in 2010! The industry's output has grown rapidly from US$ 9.4 billion in 2007 to US$ 14 billion in 2010 at 14 per cent compound annual growth rate. It is projected to reach US$ 55 billion by 2020.

The period between 2000 and 2010 has seen the top 10 companies growing from sales turnovers of ₹ 500–800 crores/year to professionally-run TNCs making hi-tech genetic drugs with turnovers ranging from ₹ 4,500 to ₹ 10,000 crores/year.

If most of these companies earlier relied on bulk drugs supplies and small exports to unregulated markets in other parts of Asia and also in Africa and formulation sales in the domestic market, the last decade saw them aggressively tapping the regulated markets of the USA and Europe. The Indian industry had filed only 3 marketing applicants to the US FDA in 1998: that number swelled to 168 in 2010.

Table 7.20

Sales Turnovers of the Top Five Indian Pharmaceutical Companies in 2000 and 2010

Name of the Company	2000 sales (₹ in crores)	2010 sales (₹ in crores)
Ranbaxy	1745	9786
Dr Reddy's Laboratories	984	8120
CIPLA	792	695
Lupin	650	5610
Sun Pharma	613	5188

Source: Author.

As for R&D, the last decade saw a 20-fold jump in spending with the R&D-to-Sales of the 'top 20' companies increasing from 1–2 per cent in 2000 to 5–10 per cent in 2010.

Table 7.20 lists the sales turnovers of the top five companies in 2010 as compared to those in 2000.

The top 20 companies made 72 overseas acquisitions over 2000–10, 50 of which were in the US and Europe. So, on every count, the Indian pharma industry is way ahead of its Chinese counterpart.

Moreover, the decade has seen the emergence and rapid growth of wholly biotechnology (BT)-based pharma companies like Biocon, Shantha Biotech and Bharat Biotech and even several of the top 10 chemistry-based pharma companies diversifying into biopharmaceuticals with focused R&D, production and marketing activities. Biocon has even become a major exporter of BT products at the world level.

Conclusions

Summing up the Chinese and Indian S&T and industrial capabilities across all the seven major industrial sectors dealt with in this chapter one comes to the conclusion that India is:

In steel:	Far ahead of China across the board, for example, technological efficiency, energy efficiency and environmental impacts, except in gross output.
In pharma:	Far ahead of China across the board.
In electrical power:	Ahead of China in most areas, except in gross output.
In renewable energy in general:	At the same level as China.
Wind power:	Ahead of China except in aggregate output in MW.
Solar power:	Behind China in most respects.
Telecom:	Behind China in most respects

8

The Growing Science and Technology Gap with China and How India Can Close It

Smita Purushottam

Everyone except India seems to be aware that the decisive contest for world power is taking place in high technology.

There are several criteria for determining a country's technological power and capacity to generate innovation. The Global Innovation Index produced by a team led by INSEAD has measured country 'innovativeness' across 80 indicators grouped under several categories. Out of 125 countries examined, India ranks 62 in their evaluation and China 29.

China's better performance in Science and Technology (S&T) indicators is a result of the specific economic and technology strategies it evolved. A holistic approach to pinpoint the causes of the S&T gap with China has therefore been adopted in this chapter. The different economic and technological paths taken by the two countries and their results and implications for the development of a sound S&T eco-system are briefly reviewed. A few recommendations on how to put India back on track to achieve fast growth, based on a more sustainable, high-tech manufacturing economy which can help India meet its domestic and external challenges, are also attempted. This chapter has, however, tried to avoid the most well-known differences which are the subject of dozens of existing commentaries; so, the analysis may still seem incomplete.

China's Economic and Technological Strategy

So What Did China Do Right?

According to Professor Justin Yifu Lin,[1] China switched from an import substituting, capital-intensive, heavy-industry development strategy, influenced by the Soviet model, to a Comparative Advantage Following (CAF) strategy of development under Deng Xiao Ping in 1978. Underlining the merits of CAF, Justin Yifu Lin stated:

> A developing country that relies on its comparative advantage to guide its choice of industry and technology will be most competitive in domestic and international markets, producing the largest possible economic surplus, accumulating the most capital possible and upgrading its endowment structure as well as its technology and industry in the fastest possible way.[2]

These policies led to what he justifiably calls the 'China Miracle', i.e., sustained growth over two decades propelling China to the status of the second biggest world economy in 2010.

China's post-1978 reforms thus reversed decades of dirigisme in that orienting the economy to exploit its natural factor endowments released the State somewhat from the burden of directing, and inevitably distorting, economic development. Development of a heavy industrial complex had required the suppression of market forces; otherwise resources would have spontaneously flown into sectors where the natural *comparative advantages* of the economy lay. China's leadership achieved these controlled outcomes only by administratively commandeering resources through central planning and bureaucratic dirigisme, a scenario with which India was also familiar, having chosen a similar path. This autarkic model was further characterized by a closed capital and current account, overvalued exchange rate, limited FDI and trade flows, and discouragement to investment in and production of consumer goods. Welfare levels, understandably, increased very slowly.

To China's credit, it woke up to its follies in 1978 when Deng Xiao Ping launched reforms by phasing in market forces, dismantled skewed and overvalued foreign exchange rates and price controls, freed up state-owned enterprises (SOEs) to participate in the market, opened up the economy albeit in a selective and calibrated manner, built world-class infrastructure starting with five Special Economic Zones (SEZs), extended preferential treatment to foreign investors, especially in SEZs, and enabled the peasant to lease his land, sell produce on the market and substantially increase his income. The government simultaneously embarked on a policy of massive investment in infrastructure, including in power generation. This led to a boom in production and the explosive growth of township and village enterprises (TVEs). By jettisoning an exclusively pro heavy-industry policy in favour of CAF, China was also able to join the 'flying geese' phenomenon which spread prosperity, investment, technology and trade flows from Japan and, later, from South Korea and Taiwan to the Southeast Asian region.

The CAF strategy is not a neo-colonial imposition aimed at keeping developing countries stuck at a lower level of aspiration and achievement in the building of a high-tech economy. The successful development of labour-intensive and presumably low-tech industries (at least in their initial stages) has demonstrably led to income growth, trade surpluses, productivity improvements including in the Ease-of-Doing-Business indexes, and to a progressive upgrade of the economy. This has ultimately laid the foundations for the development of the cherished high-tech manufacturing sector, which requires smart government intervention and State support. China has been following precisely this strategy, first focusing on deepening and diversifying its manufacturing sector, and then switching to a high-tech strategy.

India's Current Economic Crises

India's reforms, implemented a good 13 years later, followed this trajectory, but without the emphasis on building manufacturing capacity through the creation of a world-class infrastructure, an enabling business environment in which clearances were simplified and fast tracked, and agricultural reforms that created the initial pool of capital. The reforms stayed confined to the urban areas, the foreign trade and investment sector with some easing up of licencing, and were extended, a decade later, to the telecommunications and banking sectors.[3] Attempts were made at privatization and pushing PSUs to be more market-oriented, but the cumbersome bureaucratic edifice requiring permissions and 'lubrication' for setting up manufacturing enterprises continued. There was little emphasis on infrastructure creation, including power generation. India did not create the large diversified manufacturing base, as China had done so successfully, which could have employed millions of surplus rural labourers, who were condemned, instead, to turn to rural doles and temporary employment schemes.

However, for some time everything had seemed to be going right for the Indian economy, as it reaped the efficiency gains of the limited reforms undertaken. We were supposed to have reached the take-off stage in the first decade of the 21st century, similar to the East Asian miracle, on the basis of liberalization, opening up, robust savings and investment rates, productivity increases and other improvements in the business environment. The consumption and the services sector-led model of the Indian economy was compared and contrasted favourably with China's single-minded focus on production for export and under-valuation of its currency. Economists even lauded the fact that India's economic structure resembled those of the developed countries with a greater share in GDP of the services sector, and in India having, apparently, skipped the manufacturing stage altogether. This was an exaggeration as India has a robust manufacturing and engineering sector, but one which accounts for only 16 per cent of the economy and suffers from other shortcomings as will be clarified later.

But in famous and prophetic last words, Lehman Brothers had said in their 2006 report 'India: Everything To Play For': "With the right reforms, India could grow at 10% for a decade.... But that would depend upon India's continuing to make progress with structural economic reforms."

These superlative diagnoses, while deserved to some extent, stirred profound unease in some quarters. Amidst all the backslapping and self-congratulation at the World Economic Forum and other fora, a few simple truths were glossed over. How could a huge economy like India survive on services alone? Surely, a diversified manufacturing sector was equally, if not more, important for India's balanced development and for feeding its consumption-led economy? Even within the services sector, given the problems in India's higher education sector—it was clear that there were inherent limitations to the IT growth story, and therefore also to India's capacity to resolve its employment issues. Without adequate R&D and a manufacturing sector to experiment upon, India could not win dominant global market shares, launch iconic products or take advantage of the beckoning revolution in embedded technologies. The services sector was neither the answer to India's unemployment problems, nor was it going to compensate for the growing and inevitable deficit in the merchandize trade balance. India's winning streak had to be sustained through urgent structural reforms to ensure continued growth and international competitiveness. Innovation, technological progress and the related issue of broad-based manufacturing were the keys to ensuring India's international competitiveness: but for a long time they were not even part of the national discourse.

Ten years ago, while most people were busy predicting China's crash and finding terminal faultlines in its development model, I had suggested instead that:

China (1978–1997) had created an economic juggernaut based on a solid manufacturing base through policies involving State control over the economic reforms process, huge infrastructure investments, and open FDI and export oriented policies. India's first phase not only started 13 years later, but also witnessed the launching of only the easy reforms between 1991 and 1993. Neglecting infrastructure, India also did not derive full advantage from the limited liberalisation effected in this first phase. The current economic slowdown and whittling down of the manufacturing sector in India can be attributed to the neglect of core sector reforms (something China tackled early on) with a host of repercussions for sustained growth in many sectors.[4]

Now, as the limits of efficiency gains of India's curtailed reforms package are exhausted, for the first time since the 1990s serious question marks are being raised on India's economic trajectory. On 12 May 2012, the rupee depreciated further to ₹ 53.64 to the dollar, one of the lowest points reached ever. The consumption-led services-sector-based and low-tech Indian economy generated a growing fiscal and merchandize trade deficit of US$ 185 billion for 2011–12, with capital outflows aggravating the problem. The Indian Ministry of Commerce projected the deficit at an unsustainable 12–13 per cent of GDP in 2014.[5]

India is also suffering from unsustainably high inflation which makes a mockery of the claims that its economic model is benefitting the poor.

The poor today can no longer afford even basics like pulses, let alone fruit in their diet. This has frightening consequences for the health and productivity of the nation. Questions over India's success story are being raised again, with some reputed analysts going as far as to predict that India is in danger of reverting to the Hindu rate of growth if it does not tackle the "failure to export", its public finances, and the poor state of its infrastructure.[6]

Neglect of these issues has resulted in the following:

1. A growing merchandize trade deficit. According to the Ministry of Commerce, India's merchandize trade deficit will reach an unsustainable 13 per cent of GDP by 2014.

2. While the Chinese government has till recently been doing everything possible to restrain appreciation of the Renminbi, the Rupee has been falling over several months and is now known as Asia's worst performing currency. Its fall is connected to the burgeoning fiscal and trade deficits caused by imports of over 70 per cent of our machine tools requirements (the basic building blocks of industry), 90 per cent of India's telecommunications and IT hardware equipment, 100 per cent civil aerospace equipment and aircraft, 90 per cent of rail equipment[7] and 70 per cent of defence requirements.[8] We missed our chance in 1999 to insert provisions forcing telecom companies to create a manufacturing base in India.

3. And now, inevitably, this phenomenon extends to consumer goods, as the Indian manufacturing sector gets hollowed out, a term reportedly used by the National Security Council Secretariat. The fridges, air conditioners, household appliances, pens, stationary and many other items in every store are imported from China or South Korea.

4. As of now the Services sector is acting as a net drain on India's foreign exchange reserves. It comprises 55 per cent of GDP but only 35 per cent of total exports!

5. High growth in India's services sector and consumption-led economy translates into gains mostly for foreign manufacturers of aircraft, telecom equipment and other imports and education providers.

6. Except for IT and healthcare, India has not reached the 'controlling heights' in the services sector. With the right reforms India's

education sector can satisfy the country's own requirements and become a net foreign exchange earner. Today, millions of Indian students are forced to seek higher degrees in foreign universities, as India's top colleges run out of seats after ridiculous cut-offs. Educational reforms will help boost the IT sector too.

This import dependence is harming India's national welfare and national security. India has no excuse for this state of affairs. No embargoes are preventing it from meeting its requirements; for according to an IMTMA report, only a few items are under export controls, and yet India imports over 70 per cent of its machine tools requirements. By contrast, China has accumulated US$ 3 trillion in foreign exchange reserves. It became the largest exporter and the second biggest world economy. Even the US is apprehensive at the rate at which China is developing and, perhaps, even catching up with it, China's political and economic problems notwithstanding. It is clear that the Indian situation demands urgent and drastic reforms.

Our original sin was that we did not have a manufacturing or a technology policy suited to our own needs. Even after 1991, we implemented only the easy reforms, paying little attention to the reforms needed to liberate the potential in the brick and mortar space: for example, we did not eliminate the licence rentier raj. The manufacturing sector shrank as a result. India adopted a national manufacturing policy and set up a National Manufacturing Competitiveness Council (NMCC) dedicated to reverse the decline in manufacturing only last year, but it still does not have a viable national technology policy.

Part of the reason why we woke up so late may be that we were seduced by the analyses and policy prescriptions handed to us from outside. The time has indeed come to throw off these mental 'lakshmanrekhas'.

China's Technology Development Strategy

China had set its goals very high from the very beginning, and its overarching objective was to become a high-tech economy. China's technology and innovation development strategy rested on a holistic approach: national security, technology, and later (under Deng) economic development. It evolved from a top-down, techno-nationalistic approach to a more nuanced handling of various aspects of innovation. As we noted, at first China created a strong civilian manufacturing foundation based on the CAF strategy, on which tinkering and innovation of all kinds was possible. The wealth generated thereby prepared the ground for the technological upgrade of the economy. This is important as a country cannot expect to leapfrog into disruptive, radical innovation straightaway. China's multifaceted strategy to create a national innovation ecosystem comprised the following:

1. **Reverse Engineering**: China followed an aggressive strategy of reverse engineering, copying and stealing technology. There are different stages of the much derided reverse engineering phenomenon, which correspond to the level of sophistication attained by an economy. Professor Tai Ming Cheung has lucidly categorized these stages as: duplicative imitation; creative imitation; creative adaptation and/or incremental innovation; architectural innovation (which China has reached); component innovation and radical/disruptive innovation.[9] This is a legitimate strategy, if carried out within limits. Everyone has done this and India should not shy away from it.

2. **Developing Both Hard and Soft Capabilities (Professor Tai Ming Cheung):**[10] While China has still to master component and disruptive innovation—it has focused on creating both 'hard' and 'soft capabilities'—terms given by Professor Tai Ming Cheung— necessary for enabling radical indigenous innovation. Professor Tai Ming Cheung has categorised 'hard capabilities'[11] to include funding, manufacturing capabilities, laboratories, research institutes and universities. China is investing massive amounts of funding to spur innovation and technological breakthroughs and is equipping its academies, laboratories and firms with R&D capacities.

   Meanwhile, 'soft capabilities',[12] which it can be argued perhaps constitute the harder part, include relative intangibles like leadership, governance regimes, reforms, re-engineering of organizations and cultures, market reforms, sources of funding, entrepreneurial skills and resources, business environment, education and training, and reforms of institutional, governmental, educational and industrial structures. They may also include many of the parameters specified in the hugely sophisticated technique for measuring a country's Innovation Index by INSEAD. The conversion of state defence enterprises into shareholding entities referred to below would probably come under this category. Only a combination of the two, coupled with an economy's increasing sophistication, will ultimately yield disruptive innovation of the kind seen in advanced economies.

3. **Restructuring of State Enterprises**: China is actively fostering a transition from State to market-oriented enterprises, while maintaining the centrality of the State's role. The US–China Economic and Security Review Commission (USCESRC) Report also states:

> The Chinese government recognizes that national science programmes alone are not capable of sustaining the leapfrogging scientific capabilities the PRC now seeks. Although they have aided China's technological advance substantially, these programmes have not yet fostered the widespread commercialization of internationally-competitive technologies originating from Chinese R&D efforts. China's science and technology (S&T) policy now embraces the idea, conveyed in China's national plans and official speeches, of speeding up the construction of an innovation system that takes enterprises as the center, the market as guide, with commercialization and research interwoven.[13]

As per the guidelines for the Medium and Long Term National Science and Technology Development Programme (2006–20), the PRC State Council, the Five-Year Plans and other official documents and the companies are now the centre of China's national innovation system. China supports enterprises' R&D efforts through tax incentives, subsidies, investments, financial support for firms' international expansion and acquisitions, loans, procurement policies, and preferential land allotments, and not just for SOEs.[14] Some more incentives are being considered under the more recent $1.5 trillion Strategic Emerging Industries Decision. According to Professor Tai Ming Cheung,[15] 22.5 per cent of the state-owned defence firms had completed their shareholding restructuring by the end of 2007, compared with 65 per cent in the national economy.

According to the Global Times, "China is set to complete shareholding reforms of its military industries within three to five years in a bid to accelerate the process of bringing them in line with civilian businesses and to raise more funds from capital markets".[16] The Global Times report goes on to make several points that should be noted by Indian planners who are hesitating to reform the Indian PSUs, the DRDO, and other state-owned entities:

> The scheme is one of seven major tasks for defense-related science, technology and industry under the 12th Five-Year Plan (2011–15). Sources from the Ministry of Industry and Information Technology revealed that an inter-ministerial coordination mechanism would be established in 2011 to ensure the improvement of the reforms, the Shanghai Securities News (SCN) reported.
>
> The completion of the reforms is also one of five targets set by the State Council and the Central Military Commission under a guideline for the improvement of a research and production system for weapons and equipment.
>
> "The guideline has upgraded military and civilian integration to a national strategy. This is a way to revitalize those companies through actively participating in market competition," Liu Jiangping, a Chinese military expert, told the Global Times.
>
> "The enclosed structure of the military industry will be broken, the age of the 'iron rice bowl' will go, and an open, competitive and vibrant system will be established," Lin Zuoming, president of the Aviation Industry Corporation of China (AVIC), was quoted by the SCN as saying.
>
> AVIC now has nearly 200 subsidiaries (branches) and over 20 listed companies. In 2009, it was listed in Fortune 500, the first Chinese military enterprise to make the list. The enterprise is undergoing further reform in order to get listed as a single entity.
>
> "Foreign experience showed that the securitization of military assets, mergers and acquisitions are important methods of achieving rapid development in the military industry," Liu said, citing Lockheed Martin, Boeing and Northop Grumman Corporation as examples.
>
> It is necessary for Chinese military enterprises to attract capital from stock and securities markets as research and development require huge investment while their current asset-liability ratios are generally above 60 percent, which makes large-scale loan from banks or issue of bonds impossible.[17]

4. **Re-engineering Government Structures**: China has been experimenting with re-engineering government structures to coordinate and target them more precisely on the objective of technological upgradation. This is an extremely complex topic and interested readers may like to look it up in Professor Tai Ming Cheung's *Fortifying China*.[18]

5. **Leadership**: The top Chinese leadership, which is composed of technocrats, is single-mindedly directing the drive for technological upgradation. China's armed forces have been a part of this endeavour. This has been crucial for sorting out turf issues and bureaucratic opposition and ensuring effective implementation.

6. **Focused Policy Framework for Indigenization**: China has issued clear policy guidelines such as the landmark 'Guidelines for the Medium and Long Term National Science and Technology Development Programme (2006–20) of the People's Republic of China State Council'. It has launched several programmes for the promotion of the S&T sector, the earliest amongst its more modern approaches probably being the 863 programme.

7. **Civil Military Integration**: Having succeeded, at first, in creating a diversified manufacturing foundation through industry-friendly measures, China has encouraged greater integration between its civilian and defence production sectors, as proclaimed in its 16-character policy. It knows that a dual-use manufacturing base under a civil military integration (CMI) paradigm, enshrined in its main S&T development plans and its 5-year plans, can spur technological advances in both sectors. It seeks to emulate the integrated industrial structures of high-tech western economies, by essentially creating a dual-use industrial economy in which technology flows freely between the civilian and military sectors. This also enables China to assimilate technology received through civilian channels and then transmit the knowhow to the defence economy. Coupled with the massive reverse engineering that China has been doing with Soviet/Russian/Ukrainian and other military products, a significant amount of military and dual-use technology is within the reach of China's military. The Pentagon's

annual report on China's military[19] and other reports to the US Congress underline that civil–military integration, development of innovative dual-use technology and an industrial base that serves both military and civilian needs is among the highest priorities of China's leadership. President Hu, in his political report to the CCP's 17th Party Congress, stated

> We must establish sound systems of weapons and equipment research and manufacturing ... and combine military efforts with civilian support, build the armed forces through diligence and thrift, and blaze a path of development with Chinese characteristics featuring military and civilian integration.

China's defence industry has benefitted from integration with China's rapidly expanding civilian economy and science and technology sector, particularly elements that have access to foreign technology. Progress within individual defence sectors appears to be linked to the relative integration of each—through China's civilian economy—into the global production and research and development (R&D) chain.

For example, the shipbuilding and defence electronics sectors, benefitting from China's leading role in producing commercial shipping and information technologies, have witnessed the greatest progress over the last decade. Information technology companies, including Huawei, Datang, and Zhongxing, maintain close ties to the PLA and collaborate on R&D.[20]

8. **Reforms in Research Institutions, Science Academies and Educational Institutions**: China has simultaneously reformed its research institutions and science academies in order to make them more market-oriented even while it has retained the development of strategic technologies within the government sector. China has also implemented a wide range of educational reforms to increase the quantity, quality and delivery of its educational institutions.

9. **Techno-Nationalism and Techno-Globalism**: During the first decades after 1949, China followed a techno-nationalist policy focused on national security, mobilization of resources for big projects, and total state control. China's Innovation Strategy also relies on 'techno-globalism'—i.e., leveraging China's

international linkages for national benefit which the Pentagon report also mentions. Recently, China has intensified acquisitions of technology companies abroad, especially in the aviation sector. Foreign enterprises and R&D centres are proliferating in China and contributing to diffusion of technology in the country.

US Reactions to China's Successes in Technology

Admiration, Competition and Irritation

Clearly, China's reforms, and the determined push to modernize its S&T structures have had astounding results. This is recognized by the most advanced powers, led by the US where this issue is the subject of Congressional study and debate. The US reaction is complex, reflecting the dilemmas of the 21st century, where rival powers have to cooperate and compete at various levels.[21] The US reaction is, therefore, composed of 'acknowledgement' of China's achievements, which should put paid to our own sceptics in India. Determination to compete coupled with a mixture of admiration and irritation is part of America's reaction. Also, there is genuine concern over national security implications! Despite all this, neither the US nor the Europeans, who have made the most noise regarding unfair Chinese practices in technology, are in a position to curtail their commercial engagement with China.

Thus, the US is sitting up and taking note of China's technological achievements. With a 10 per cent share of global merchandize exports, even if allowances are made for processing trade, China's global competitiveness is clearly established, as is proven by its export dominance in several high-tech sectors. The ASBM and the new jet with stealth capabilities has shocked everyone out of their complacency regarding China's new capabilities. The report 'China's Program for Science and Technology Modernization', prepared for the USCESRC (January 2011), shows a clear-eyed acknowledgement of China's growing technological prowess:

• China is no longer just the world's workshop. Manned space ventures, electric cars, and the world's fastest supercomputer all make clear: the People's Republic of China (PRC) is *ascendant* in science and technology.

According to Secretary of Energy Steven Chu, speaking in late 2010, China's recent technological successes constitute a new 'Sputnik moment' for the United States.

- With China poised to be a leader in clean energy and transportation technologies, Secretary Chu was suggesting a technological challenge on a level that ought to shock the American psyche.
- China's low-emission coal energy plants, third and fourth generation nuclear reactors, high-voltage transmission lines, alternative-energy vehicles, solar and wind energy devices, and high-speed trains, are all either more advanced than those in the United States, or provide serious competition to American technologies.
- The transformation in Chinese technological capabilities is not only apparent in the clean energy and transportation fields. China's high-tech industries have made steady progress in telecommunications and information technology (IT).
- Significant budgetary commitments for research in nanotechnology, new materials, and other cutting edge scientific fields have allowed China to play a leading role in the next generation of important discoveries.
- And advanced military weapons systems (including recently deployed anti-ship ballistic missiles and a new fighter jet prototype with stealth characteristics), have benefited from advances in the PRC's defence industries and in China's civilian technology base.[22]

That China's technological progress indeed constitutes a new Sputnik moment for the United States had found mention even in President Obama's 2011 State of the Union Address. Referring to the race between the superpowers to reach the moon sparked by the Soviet Union's pioneering successes in space technology, the President said:

Half a century ago, when the Soviets beat us into space with the launch of a satellite called Sputnik, we had no idea how we'd beat them to the moon. The science wasn't there yet. NASA didn't even exist. But after investing in better research and education, we didn't just surpass the Soviets; we unleashed a wave of innovation that created new industries and millions of new jobs. This is our generation's Sputnik moment.

It is perhaps interesting that the challenge to the US, once again, comes from another at least nominally Communist country, and that everyone except India seems to be aware that the decisive contest for world power is taking place in high technology!

At the same time, the USCESRC report also gives a measure of American irritation with China's outright copying of technology and thus goes on to criticise some of China's more blatant techniques for capturing technology:

- While China remains a long way off from challenging the US for leadership, the trajectories of the two countries warrant serious attention. Already, the world has seen China's scientific efforts become a bone of contention and suspicion as its advances are directed into areas of competition with other nations. After all, noted PRC President Hu Jintao in a 2010 speech, "a nation's technological competitiveness determines its place and future in international competition".
- Techno-nationalist practices have at times undermined the mutually beneficial basis for the exchange of knowledge and goods across borders. Instances in which China has created an unfair playing field for foreign companies in the high-tech sphere, or stolen foreign technologies in order to 'free ride' on the advances of others, have stimulated fears that foreign nations are not only failing to obtain an adequate return on their significant investments in Chinese science, but that their efforts will come back to harm them in the future—if they have not already. For there may be more 'Sputnik moments' of the kind described by Secretary Chu, moments in which China's technological achievements suddenly awaken foreign nations and enterprises to the fact that the old paradigms guiding their interactions with the PRC in science and technology are no longer applicable.[23]

China has been criticized by many others also for strip-mining foreign products for their technological content. The Russians criticize China for copying their military technologies; German CEOs complained over IPR infringements during Chancellor Merkel's visits to China; the Chancellor herself has publicly attacked such practices and many firms have levelled charges at their Chinese partners.

In 2010, a rule mandating compulsory domestic innovation content to make products eligible for preferential treatment in Chinese government tenders led to an outcry against China. NASSCOM and European, American, Canadian, Korean and Japanese business associations joined hands to write a letter to the Chinese authorities, alleging discrimination and requesting China to withdraw the provisions. The European Chamber of Commerce echoed these demands. The US Chamber of Commerce castigated China for "technology theft on a scale the world has never seen before".[24] All this as if they had just discovered the practice! The Cox Committee report of 1999 had documented in great detail how China had acquired US aerospace, warhead and satellite technologies clandestinely.

Great pressure was brought to bear on China, and now also on India, to adhere to the WTO agreement on Government Procurement to ensure that innovative domestic firms do not get any preferences in domestic government orders. Such a rule makes eminent sense for a country desiring to incentivize the development and incubation of indigenous technology. India should resist such binding measures and commitments, as China has done, in order to, at least, keep its options open.

So, given the hue and cry against China, what could one have expected the result to be? Embargoes against dealing with China in the high-tech field? Disengagement from China and a rethink of China-focused strategies by Western high-tech firms? Not really. The West does maintain an arms embargo against China and some restrictions on export of military technologies. But the increasing cooperation in the aerospace and other high-tech sectors has meant that there have been significant transfers of technology to China over the years. The fact is that the overriding attractions of the Chinese market have systematically neutered the desire for retaliatory action by the west. China has also found an ingenious method to circumvent export controls on high-tech products through the civil-military integration (CMI) paradigm referred to above.

The case of the General Electric (GE) Company is interesting.[25] A report in the *New York Times* of 17 January 2011 says:

> no Western company has been more aggressive in helping China pursue that dream than one of the aviation industry's biggest suppliers of jet engines and airplane technology ... G.E., in partnership with a state-owned Chinese company, will be sharing its most sophisticated airplane electronics ... because the commercial aircraft market in China is expected to generate sales of more than $400 billion over the next two decades, it is not a party the company is willing to miss. The G.E. avionics joint venture ... appears to be the deepest relationship yet and involves sharing the most confidential technology. And G.E.'s partner, Avic, also supplies China's military aircraft and weapons systems. To address American government security concerns, the joint venture in Shanghai will occupy separate offices and be equipped with computer systems that cannot pass data to computers in Avic's military division, G.E. executives say. And anyone working in the joint venture must wait two years before they can work on military projects at Avic.[26]

A cast-iron, fool-proof safeguard indeed!

This is the same GE whose CEO, Jeffrey Immelt, was quoted widely in 2010 for roundly criticizing the mercantilist Chinese approach to business.

Concern over National Security Implications

China's rise as a technology power is also prompting concern over its implications for American power, as the USCESRC report makes clear:

- China's emergence as a major force in science and technology has profound implications for the United States....
- China's continued advance in science and technology may significantly alter the distribution of global economic, political and military power to the disadvantage of the United States....
- Gains in China's technical capabilities also support military programs that threaten the interests of the United States and its allies.[27]

Technology thus has a clear national security dimension.

Lessons for India

A legitimate question arises when one transitions to examining the relevance of the Chinese model for India: is the Chinese innovation model relevant to India? More importantly, can it be applied in the Indian context, given the silo-based functioning not only between, but also within ministries? There are many sceptics in India about the Chinese model. The three most common criticisms are that (*a*) it is a top-down, state-driven model, (*b*) the quality of products is poor, and, of course, (*c*) copycatting. All these are valid to some extent. That is why it is important that we must also learn from other models and keep an open mind. But these others lack the comparable scale of China for India.

The counter arguments to the above are that China, which started with the same disabilities like central planning that India started with and is still struggling with the legacy, is trying to graduate away from the top-down, bureaucratic approach to one which places greater emphasis on the market and is driven by enterprises and universities and not only by the State. This is extremely relevant to India, given the stepmotherly treatment still given to private enterprise. Moreover, the reforms, policy initiatives and strategies in multiple sectors initiated by China contain crucial lessons and one key message for India—the need to pursue a coordinated and single-minded strategy to achieve goals. As for the statist argument, even the US government has, from the beginning, supported

S&T. The State has, in fact, a big role to play in any innovation model. The Chinese State has devoted a huge amount of funds to the R&D sector. At the same time, the experience of another country can never be copied wholesale, and as I have observed elsewhere, in any case copycats do not catch mice. We have to preserve the ability to think for ourselves and bring about change suited to our requirements.

The lessons for India are simple:

Lesson No. 1:

The necessity of creating an internationally competitive, broad-based, high-tech manufacturing economy is staring at us in the face. India must impart greater dynamism to economic reform and grow the manufacturing sector, both in traditional and high-tech areas. Even as China vacates the traditional sectors, countries like Vietnam, Cambodia and Bangladesh have captured market shares in textiles, shoes and other competitive products, while India with its teeming millions has been unable to compete.[28] India cannot have cutting-edge technology without a diversified manufacturing base so this is an area where all national effort needs to be focused. India has finally made a good though belated beginning by adopting a national manufacturing policy and setting up a National Manufacturing Competitiveness Council (NMCC).

Lesson No 2:

India must ensure that no measure is taken to damage the viability of Indian industry, even at the cost of appearing protectionist. If China has risked this, so can we. For example, pressure to conclude free trade deals that disadvantage our engineering goods, automotive and pharmaceuticals sectors must not be given in to. India has created only a few viable manufacturing sectors and we cannot afford to give an inch in these. This is not to suggest autarky, but to ensure that the primacy of Indian industry and manufacturing is enabled with supportive policies.

Similarly, any pressure to join the WTO agreement on government procurement, which damages our manufacturing sector and disadvantages Indian industry in government procurement, has to be resisted.

Lesson No. 3:

India should not shy away from Reverse Engineering. There is no shame in it—everyone has been there, done that. The consequences have to be intelligently managed. After all, the success of India's pharmaceutical sector, now a target of foreign acquisitions, is due to reverse engineering. We need to additionally ask ourselves the question—why have we allowed a liberal acquisitions regime in the pharmaceuticals sector? Foreign companies would jump at the chance to eliminate Indian competitors and moreover, generics are reportedly going to drive profits for the foreseeable future. It is essential to consolidate the lead we have gained in this sector, rather than sell it off.

Lesson No. 4:

Both developing hard and soft capabilities are needed and must be developed. Experience shows that building soft capabilities is actually the harder part! It is not important just to throw resources at the problem and create newer and newer institutions, committees, commissions, etc. It is more important to reform the way the system works.

Lesson No. 5:

Leadership is, actually, a prerequisite for all other reforms! China's leadership is prioritizing and directing the drive to make China the leading S&T power. China pursues a holistic strategy though all government departments, such as the Ministries, the State Council and the Armed Forces, through policies like offsets, manufacturing, TOT and FDI—to ensure domestic technological upgradation. The re-engineering of government structures, resolving turf battles and bureaucratic infighting, and ensuring implementation become possible when there is top-down direction to the process.

Leadership is crucial, as the kind of measures that we want to see implemented need intervention at the highest levels of government. India's leadership has to take the hard decisions required to develop an indigenous innovation ecosystem and a high-tech economy.

Lesson No. 6:

Clarity of objective is of critical importance. India needs focused policy guidelines for indigenization. China has several guidelines and programmes—the 863 programme (and supportive higher education national programmes), the landmark 'Guidelines for the Medium- and Long-Term National Science and Technology Development Programme (2006–20) of the People's Republic of China State Council' and other state-funded programmes (SEID). We have no set of practical guidelines, which can be enforced through the system and even the DPPs 2011 are overly long, complex and confusing. Given the utter lack of direction and coordination, this makes the job of various departments even more confusing as no one has even been told that indigenization should be the country's top priority. The injunction in the January 2011 Defence Production Policy to give priority to indigenous defence production must be more rigorously translated into concrete projects.

Lesson No. 7:

The Civil Military Integration (CMI) paradigm and the synergies it bestows for technology development is extremely useful for India. It leads to development of a high-tech, dual-use sector which is beneficial for the standard of living and national welfare. Indeed a sophisticated defence economy cannot thrive without a sophisticated economy-wide supply chain. This has important implications for creating a national Offset Policy and also a more broad-based *defence* offset policy as it resolves the debate on direct and indirect offsets. The Chinese Government is hastening the conversion of state-owned defence enterprises in recognition of the need to build maximum synergies between the two sectors.

Lesson No. 8:

Re-engineering of institutions, especially MOD, should be given high priority.

An excellent description of how China re-engineered institutions to enable technology development is contained in Professor Tai Ming Cheung's path-breaking book *Fortifying China*. Creating optimal, streamlined structures for achieving this goal is something India needs to do forthwith, especially in the MOD.

Lesson No. 9:

India needs to seize the opportunities created by favourable circumstances and launch a defence industrialization strategy.

TIME TO ACT IS NOW as there are four brilliant openings to build an indigenous high technology defence innovation base:

1. The spate of defence equipment orders which has created a unique offset opportunity.

2. Defence Production and Defence Procurement Policy reforms in 2011 prioritizing indigenization.

3. A greater willingness to partner with India and easing of restrictions on technology transfers by the US and other Western countries to Indian entities, reflecting the changing geostrategic realities, and laying the ground for the announcement of a host of agreements with significant scope for technology transfers between Indian and foreign partners. It is essential that this window of opportunity be fully availed of and high-tech production be launched in India. Over time, it will encourage development of Indian capabilities. However, India will have to guard against the possibility of succumbing to greater dependence on Western equipment inflows.

4. The global arms market is slowing down; India is in an excellent position to leverage its vast market.

A plan, which subsumes all of these lessons, now exists. On 14 July 2011, I launched the open High-Tech Forum on Defence Innovation to propose a policy and vision for India's scientific and technological advancement. I called it VISTAS, or Vision for an Integrated Science & Technology Advancement Strategy. The report of the forum was circulated to NSA, Defence and Defence Production Secretaries, Foreign Secretary, Commerce Secretary, Secretary Education, Planning Commission, the Armed Forces and Industry Associations, key members of the NSAB and the newly-constituted Task Force on Self-Reliance in Defence Production.[29] The second meeting of the forum was held on 16 December in New Delhi and was chaired by the Military Advisor of NSCS. We also created a high-tech Google group which is also an open platform, not owned by anyone, which all concerned citizens can join by sending an email to smitapurushottam@gmail.com.

If India is to face the domestic and external challenges in the 21st century and realize its potential as one of the major poles in the international system, it has to move fast to address the concerns and implement the recommendations suggested in VISTAS to create a broad-based manufacturing sector, and a viable high-tech dual use sector. Without necessary reforms and changes, India will not be able to take its place as an advanced technological nation, and that will affect its autonomy in conducting foreign policy, but most importantly, also adversely affect the overall welfare of its citizens.

Notes and References

1. Chinese economist, former Director of the China Centre for Economic Research at Beijing University, and now Chief Economist at the World Bank.
2. Justin Yifu Lin, *Economic Development and Transition: Thought, Strategy, and Viability* (New York: Cambridge University Press, 2009).
3. Smita Purushottam, *Can India Overtake China?* (Harvard: Weatherhead Center for International Affairs, Harvard University, 2001).
4. Smita Purushottam, *Can India Overtake China?*
5. This was pointed out in the chapter 'Technology as the Key to Power' in *Present-Day China: A Contemporary Assessment*. ed. M. Rasgotra (New Delhi: Academic Foundation, 2012): 33–54.
6. Jonathan Anderson, 'And the Three Reasons India Will Fade Away', Emerging Advisors Group, 28 April 2012.
7. Contrast this with China's success in domestic high-speed rail production and the efficiency with which it runs its modern, 21st century railway system.
8. This last figure is not even backed by reliable research, for example, the Government's CAG report states that the import component of the so-called indigenous Dhruv Light Air Helicopter is 90 per cent and the Brahmos is also preponderantly made up of imported components.
9. Tai Ming Cheung, 'The Chinese Defense Economy's Long March from Imitation to Innovation', *Journal of Strategic Studies* 34, no. 3 (17 June 2011): 325–54.
10. Ibid.
11. Ibid.
12. Ibid.
13. 'China's Program for Science and Technology Modernization: Implications for American Competitiveness', prepared for The U.S.-China Economic and Security Review Commission. The information in this report is current as of January 2011. http://www.uscc.gov/researchpapers/2011/USCC_REPORT_China's_Program_forScience_and_Technology_Modernization.pdf
14. Ibid.
15. Tai Ming Cheung, 'The Chinese Defense Economy's Long March'.
16. Deng Jingyin, 'Military Enterprises to Face the Market', *Global Times* (7 January 2011), http://www.globaltimes.cn/NEWS/tabid/99/ID/661475/Military-enterprises-to-face-the-market.aspx. There are many other such Chinese media reports.
17. Ibid.

18. Tai Ming Cheung, *Fortifying China: The Struggle to Build a Modern Defense Economy* (New York: Cornell University Press, 2009).

19. 'Military and Security Developments Involving the People's Republic of China, 2010', Office of the Secretary of Defense.

20. Ibid.

21. Zorawar Daulet Singh, 'Thinking about an Indian Grand Strategy', *Strategic Analysis* 35, no. 1 (January 2011): 52–70.

22. 'China's Program for Science and Technology Modernization: Implications for American Competitiveness'.

23. Ibid.

24. James McGregor, 'China's Drive for "Indigenous Innovation": A Web of Industrial Policies', Global Regulatory Cooperation Project, US Chamber of Commerce, http://www.apcoworldwide.com/content/PDFs/Chinas_Drive_for_Indigenous_Innovation.pdf

25. David Barboza, Christopher Drew and Steve Lohr, 'G.E. to Share Jet Technology with China in New Joint Venture', *New York Times* (17 January 2011).

26. Ibid.

27. 'China's Program for Science and Technology Modernization: Implications for American Competitiveness'.

28. Jonathan Anderson, in 'And the Three Reasons Why India Will Fade Away' states:

> Forget about comparisons with China; as you can see, there are none. India can't keep up with much smaller players like Vietnam and Indonesia—and even Bangladesh and Cambodia are slowly overtaking. In fact, and these five low-income Asian economies India is the only one with falling low-end manufacturing market share over the last half-decade.

29. Smita Purushottam, 'Report of the Forum on High-Tech Defence Innovation', http://www.idsa.in/system/files/Report_HiTechDefInnovation.pdf

About the Editor and Contributors

Editor

Maharajakrishna Rasgotra was India's Foreign Secretary from 1982 to 1985. In a long career in India's Foreign Service which ended in 1990, he filled several high posts in India and abroad with distinction. His ambassadorial career from 1967 to 1990 took him to Morocco and Tunisia, the USA, the UK, Nepal, the Netherlands, France and UNESCO and finally, again, to the UK as India's High Commissioner. His tenure as Foreign Secretary from 1982 to 1985 was marked by a renewal in Indo-American relations, sustained negotiations with Pakistan and an opening to China.

After retirement from government service, Mr Rasgotra was Honorary Visiting Professor at the Jawaharlal Nehru University, Regent's Professor at the University of California, Los Angeles, President of the Delhi College of Arts and Commerce, member of the UN's Disarmament Advisory Council and Chairman of government of India's National Security Advisory Board.

Contributors

N. Balakrishnan is currently Associate Director of the Indian Institute of Science and a Professor at the Department of Aerospace Engineering and at the Supercomputer Education and Research Centre. His areas of research include Numerical Electromagnetics, High Performance Computing and Networks, Polarimetric Radars and Aerospace Electronic Systems, Information Security and Digital Library. He was the NRC Senior Resident Research Associate at the National Severe Storms Laboratory,

Norman, Oklahoma, USA, Visiting Research Scientist at the University of Oklahoma in 1990, and the Colorado State University in 1991. He was Visiting Professor at the Carnegie Mellon University from 2000 till 2006. He is an Honorary Professor at the Jawaharlal Nehru Centre for Advanced Scientific Research, Bangalore.

V.P. Kharbanda, a senior Indian scientist, till his sudden demise in January 2012, was head of the Research Planning and Monitoring Division at the National Institute of Science, Technology and Development Studies, New Delhi. He had been engaged in Science, Technology and Society policy studies on India and China since 1980, with a particular emphasis on relations between the State, the Academia and the Industry in association with several bilateral international programmes with institutions under the Chinese Academy of Sciences (CAS), the National Science Foundation of China (NSFC) and the State Science and Technology Commission, China.

Roddam Narasimha is India's leading aerospace scientist and fluid dynamicist. Early in his career he held various positions in the Department of Aerospace Engineering at the Indian Institute of Science (IISc), and founded the Centre for Atmospheric (now also Oceanic) Sciences, which he headed from 1982 to 1989. He was the Director of the National Aerospace Laboratories (NAL) from 1984 to 1993, Director of the National Institute of Advanced Studies from 1997 to 2004, and Chairman of the Engineering Mechanics Unit at the Jawaharlal Nehru Centre for Advance Scientific Research (JNCASR) at Bangalore till 2010. He is currently Honorary Professor at JNCASR. He held the Pratt & Whitney Chair in Science and Engineering at the University of Hyderabad from 2006 till 2010. For many years he held a visiting appointment at Caltech as Clark B. Millikan Professor or Sherman Fairchild distinguished scholar. He also held visiting positions at Cambridge (Nehru Professor), NASA Langley, University of Strathclyde, University of Brussels and Adelaide University. Professor Narasimha has twice served on India's National Security Advisory Board. He is currently a member of the Prime Minister's Scientific Advisory Council and of the Space Commission. He chairs the Joint Scientific Working Group for the Indo-French atmospheric research satellite Megha Tropiques.

Ashok Parthasarathi, a reputed physicist and electronics engineer, researched in Radio Astronomy at the Cambridge University with Nobel Laureate Martin Ryle. Subsequently, as Carnegie Fellow at MIT and Harvard, he had done pioneering work on the then globally virgin field of science and technology policy. From 1970 to 1975, he was Special Assistant for S&T of the late Prime Minister Indira Gandhi. In that high position, and subsequently as Secretary to Government in Electronics, Scientific and Industrial Research and New and Renewable Energy, he made important contributions in the formulation of national policy and strategy for advancement in those key sectors of Development and National Security, Space and Atomic Energy.

After retiring from Government in 2000, Professor Parthasarathi was invited by the Jawaharlal Nehru University for Chair and Professorship of the Centre for Studies in Science Policy. He built up the Centre as a leading institution in Science Policy Studies in Asia. He was consultant to leading UN agencies and has prepared several path-breaking reports for them on various aspects of science and technology policy, planning and management.

Smita Purushottam is a distinguished member of the Indian Foreign Service, known for her intellectual eminence and scholarship. At present, she is Ambassador of India to Venezuela. Earlier she had served in senior posts in India's diplomatic missions in Berlin, London, Beijing, Brussels and Moscow. In-between foreign postings, she served with distinction on the Faculties of the Institute of Defence Studies and Analysis, the Indian Foreign Service Institute and the IDS HQ in the Indian Ministry of Defence. Ambassador Purushottam spent a year as Fellow at the Harvard University during which period she prepared a paper, 'Can India Overtake China?' comparing the reform experiences of the two countries and outlining the need for a broad-based manufacturing sector and other reforms in India.

R. Rajaraman is an Emeritus Professor of Theoretical Physics at the Jawaharlal Nehru University, New Delhi. He is Co-Chair of the International Panel on Fissile Materials and Vice President of the Indian National Science Academy. He has also been doing research and teaching in physics at the Cornell University, the Institute for Advanced Study at Princeton, the Harvard University, M.I.T, the Stanford University, CERN, Delhi University and the Indian Institute of Science. He has been working on important issues of nuclear policy (both military and civilian), early warning systems, missile defence, battlefield nuclear weapons and fissile materials.

V.S. Ramamurthy is a well-known Indian nuclear scientist with a broad range of contributions from basic research to science administration. Professor Ramamurthy started his career in the Bhabha Atomic Research Centre, Mumbai in the year 1963. His scientific work, both experimental and theoretical, covers many important areas of nuclear fission and heavy ion reaction mechanisms, statistical and thermodynamic properties of nuclei, physics of atomic and molecular clusters and low energy accelerator applications. During the period 1995–2006, he was Secretary to the Government of India, Department of Science & Technology. For nearly a decade he was Chairman of the IAEA Standing Advisory Group on Nuclear Applications. He was also Chairman, Board of Governors, Indian Institute of Technology, Delhi and a highly reputed Member of the National Security Advisory Board.

Prof. Ramamurthy is continuing research in Nuclear Physics in the Inter-University Accelerator Centre, New Delhi, and is actively involved in human resource development in all aspects of nuclear research and applications. Currently, he is the Director of the National Institute of Advanced Studies in Bangalore.

U.R. Rao, former Chairman of Indian Space Research Organisation, has made extensive contributions to the development and application of space technology in India. Starting with 'Aryabhata', India's first satellite, Professor Rao was responsible for the development of INSAT and IRS series of communication and remote-sensing satellites and ASLV and PSLV launch vehicles. He was Chairman of the UN Committee On Peaceful Uses of Outer Space during 1997–2000 and President of UNISPACE-III Conference. The Space News Magazine listed Professor Rao among the top 10 International Space Personalities. Professor Rao has received numerous national and international honours and awards. Presently, he is the Chairman of the Governing Council of Physical Research Laboratory, Ahmedabad, the Karnataka Science and Technology Academy, Governing Council of Indian Institute of Tropical Meteorology, Pune and the National Centre for Antarctic and Ocean Research, Goa.

Index